THE MONTHLY SKY GUIDE

THE MONTHLY SKY GUIDE

Ian Ridpath and Wil Tirion

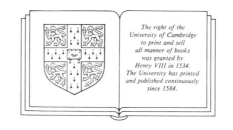

The right of the
University of Cambridge
to print and sell
all manner of books
was granted by
Henry VIII in 1534.
The University has printed
and published continuously
since 1584.

CAMBRIDGE UNIVERSITY PRESS

Cambridge
New York   New Rochelle   Melbourne   Sydney

Published by the Press Syndicate of the University of Cambridge
The Pitt Building, Trumpington Street, Cambridge CB2 1RP
32 East 57th Street, New York, NY 10022, USA
10 Stamford Road, Oakleigh, Melbourne 3166, Australia

First published 1987

Printed in Great Britain by
David Green Printers, Kettering

British Library cataloguing in publication data
Ridpath, Ian
The monthly sky guide.
1. Stars — Identification
I. Title   II. Tirion, Wil
523.8   QB801

Library of Congress cataloguing in publication data
Ridpath, Ian
The monthly sky guide.
1. Astronomy — Observers' manuals. I. Tirion, Wil.
II. Title.
QB63.R525   1987   523.8'9   87-6660

ISBN 0 521 33921 9

HO

Contents

For not in vain we watch the constellations,
Their risings and their settings, not in vain
The fourfold seasons of the balanced year.

Teach me to know the paths of the stars in heaven,
The eclipses of the Sun and the Moon's travails

INTRODUCTION

Stars are scattered across the night sky like sequins on velvet. Over 2000 of them are visible to the unaided eye at any one time under the clearest conditions, but most are faint and insignificant. Only a few hundred stars are bright enough to be prominent to the naked eye, and these are plotted on the monthly sky maps in this book. The brightest stars of all act as signposts to the rest of the sky, as shown on pages 14–15. It is a welcome fact that you need to know only a few dozen stars to find your way around the sky with confidence. This book will introduce you to the stars month by month, without need for optical aid, so that you become familiar with the sky throughout the year.

What is a star?

Every star is a sun, a blazing ball of gas like our own Sun, but so far away that they appear as mere points of light in even the most powerful telescopes. At the centre of each star is an immense natural nuclear reactor, which produces the energy that makes the star shine. Stars can shine uninterrupted for billions of years before they finally fade away.

Many stars have noticeable colours – for example, certain stars are reddish-orange, such as Antares, Betelgeuse and Aldebaran – and the colour of a star tells us its temperature. Orange and red stars have surfaces that are not as hot as that of the Sun, which is yellow-white. The hottest stars of all appear blue-white, notably Rigel, Spica and Vega. Star colours are more distinct when viewed through binoculars or telescopes.

The familiar twinkling of stars is nothing to do with the stars themselves. It is caused by currents of air in the Earth's atmosphere, which produce an effect similar to a heat haze. Stars close to the horizon twinkle the most because we see them through the thickest layer of atmosphere (see diagram). Bright stars often flash colourfully from red to blue as they twinkle; these colours are due to the star's light being broken up by the atmosphere.

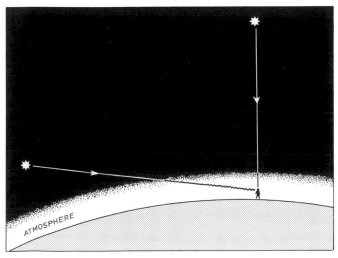

Stars near the horizon seem to twinkle, because their light passes through more of the Earth's atmosphere than light from stars overhead.

What is a planet?

Nine planets, including the Earth, orbit the Sun. The basic difference between a star and a planet is that stars give out their own light but planets do not. Planets shine in the sky because they reflect the light of the Sun. Planets can consist mostly of rock, like our Earth, or they can be composed of gas, as are Jupiter and Saturn.

Planets are always on the move, so they cannot be shown on the maps in this book. The three outermost planets – Uranus, Neptune and Pluto – are too faint to be seen with the naked eye. The innermost planet, Mercury, keeps so close to the Sun that it is perpetually engulfed in twilight. So there are only four planets that are prominent in the night sky: Venus, Mars, Jupiter and Saturn. The positions of these four planets each month for a five-year period are given in the monthly notes in this book. The planets appear close to the plane of the ecliptic, the Sun's yearly path around the sky, shown as a dotted line on the maps in this book.

The brightest planet is Venus, for two reasons: it comes closer to the Earth than any other planet, and it is entirely shrouded by highly reflective clouds. Venus is popularly termed the morning or evening 'star', seen shining brilliantly in twilight before the Sun rises or after it has set. As Venus orbits the Sun it goes through phases like those of the Moon, noticeable through small telescopes and binoculars.

The second brightest planet as seen from Earth is Jupiter, the largest planet of the solar system. Binoculars reveal its rounded disk and its four brightest moons. Mars, when closest to us, appears as a bright red star, but it is too small to show much detail through small telescopes. Saturn at its closest appears like a bright star, and binoculars just show the outline of the rings that girdle its equator.

It is often said that planets do not twinkle, but this is not entirely correct. Some twinkling of planets can be seen on particularly unsteady nights, but since planets are not point sources they certainly twinkle far less than stars.

What is a constellation?

About 4500 years ago, people of the eastern Mediterranean began to divide the sky into easily recognizable patterns, to which they gave the names of their gods, heroes and fabled animals. Such star patterns are known as constellations. They were useful to seamen for navigation and to farmers who wanted to tell the time of night or the season of the year. By the time of the Greek astronomer Ptolemy in 150 AD, 48 constellations were recognized.

Since then, various astronomers have introduced new figures to fill the gaps between the existing ancient constellations. Many of the new groups lie in the far southern part of the sky that was invisible to the Greeks. Some of the newly invented constellations have since been abandoned, others have been amalgamated, and still others have had their names or boundaries changed. This haphazard process has left a total of 88 constellations, of all shapes and sizes, covering the entire sky like pieces of a jigsaw. They are all given Latin names. The constellation names and boundaries are laid down by the International Astronomical Union, astronomy's governing body.

The stars in a constellation usually are unrelated, lying at widely differing distances from us and from each other (see diagram), so the patterns they form in the sky are entirely accidental. One fact which dismays beginners is that few constellations bear any resemblance to the objects they are supposed to represent. Constellations are best thought of not as pictures in the sky but as a convenient way of locating celestial objects.

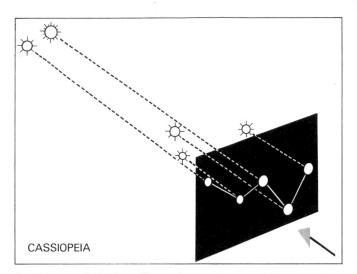

Stars in a constellation lie at different distances from us, as in the case of Cassiopeia.

How did the stars get their names?

Most bright stars, and several not-so-bright ones, have strange-sounding names. Other stars are known merely by letters and numbers. These designations arose in various ways, as follows.

A number of star names date back to Greek and Roman times. For example, the name of the brightest star in the sky, Sirius, comes from the Greek word for sparkling or scorching, in reference to its brilliance. The name of another bright star, Antares, is Greek for 'rival of Mars', on account of its strong red colour, similar to that of the planet Mars. The brightest star in the constellation Virgo is named Spica, from the Latin meaning 'ear of corn', which the harvest goddess Virgo is pictured holding.

But most of our star names are Arabic in origin, and were introduced into Europe in the Middle Ages through the Arab conquest of Spain. For example, Aldebaran is Arabic for 'the follower', from the fact that it follows the star cluster known as the Pleiades across the sky. Fomalhaut comes from the Arabic meaning 'mouth of the fish', from its position in the constellation Piscis Austrinus, the Southern Fish. Betelgeuse is a corruption of the Arabic *yad al-jawza*, meaning 'the hand of Orion'. Its name is often mistranslated as 'armpit of the central one'.

In all, several hundred stars have proper names, but only a few dozen names are commonly used by astronomers. Usually, astronomers refer to stars by their Greek letters, assigned in 1603 by the German astronomer Johann Bayer; hence these designations are known as Bayer letters. On this system, Betelgeuse is Alpha (α) Orionis, meaning the star Alpha in the constellation of Orion. Another system of labelling stars is by their numbers in a star catalogue compiled by the English

astronomer John Flamsteed. These are known as Flamsteed numbers, and they are applied to fainter stars that do not have Bayer letters, such as 61 Cygni. (Note that the genitive case of a constellation's name is always used with Bayer letters and Flamsteed numbers). Stars too faint to be included in these systems, or stars with particular characteristics, are referred to by the numbers assigned to them in a variety of specialized catalogues.

How far are the stars?

So remote are the stars that their distances are measured not in kilometres or miles but in the time that light takes to travel from them to us. Light has the fastest speed in the Universe, 300,000 km/sec (186,000 mile/sec). It takes just over 1 second to cross the gap from the Moon to the Earth, 8.3 minutes to reach us from the Sun, and 4.3 years to reach the Earth from the nearest star, Alpha Centauri. Hence Alpha Centauri is said to be 4.3 light years away.

A light year is equivalent to 9.5 million million km (5.9 million million miles), so that in everyday units Alpha Centauri is about 40 million million km (25 million million miles) away. Even our fastest space probes would take about 80,000 years to get to Alpha Centauri, so there is no hope of exploring the stars just yet.

Most of the stars visible to the naked eye lie dozens or hundreds of light years away. It is startling to think that the light entering our eyes at night left those stars so long ago. The most distant stars that can be seen by the naked eye are over 1000 light years away; examples are Deneb in the constellation Cygnus, and several of the stars in Orion. Only the most luminous stars, those that blaze more brightly than 50,000 Suns, are visible to the naked eye over such vast distances. At the other end of the scale, the feeblest stars emit less than a thousandth the light of the Sun, and even the closest of them cannot be seen without a telescope.

How bright are the stars?

The stars visible to the naked eye range more than a thousandfold in brightness, from the most brilliant one, Sirius, to those that can only just be glimpsed on the darkest of nights. Astronomers term a star's brightness its magnitude. The magnitude system is one of the odder conventions of astronomy.

Naked-eye stars are ranked in six magnitude classes, from first magnitude (the brightest) to sixth magnitude (the faintest). A difference of five magnitudes is defined as equalling a brightness difference of exactly 100 times. Hence a step of one magnitude corresponds to a difference of about 2.5 times in brightness. A difference of two magnitudes corresponds to a brightness difference of $2.5 \times 2.5 = 6.3$ times. Three magnitudes equals a brightness difference of $2.5 \times 2.5 \times 2.5 = 16$ times, and so on.

A star 2.5 times brighter than magnitude 1.0 is said to be of magnitude 0. Objects brighter still are assigned negative magnitudes. Sirius, the brightest star in the sky, has a magnitude of −1.46.

The magnitude system can be extended indefinitely to take account of the brightest and the faintest objects. For instance, the Sun has a magnitude of −27. Objects fainter than sixth magnitude are classified in succession as seventh magnitude, eighth magnitude, and so on. The faintest objects that can be detected by telescopes on Earth are about magnitude 25.

What are double stars?

Most stars are not single, as they appear to the naked eye, but have one or more companion stars, a number of which can be seen through small telescopes, and some with binoculars. The visibility of the companion star depends on its brightness and its distance from the primary star – the closer the stars are together, the larger the aperture of telescope that is needed to separate them.

Usually, the members of a star family all lie at the same distance from us and orbit each other like the planets around the Sun. A pair of stars genuinely related in this way is known as a binary. But in a smaller number of cases one star simply lies in the background of another, at a considerably different distance from us. Such a chance alignment of two unrelated stars is termed an optical double.

One of the attractions of double and multiple stars is the various combinations of star colours that are possible. For example, one of the most beautiful double stars is Albireo in the constellation of Cygnus, which consists of amber and green stars, looking like a celestial traffic light. Other impressive double and multiple stars are mentioned in the constellation notes.

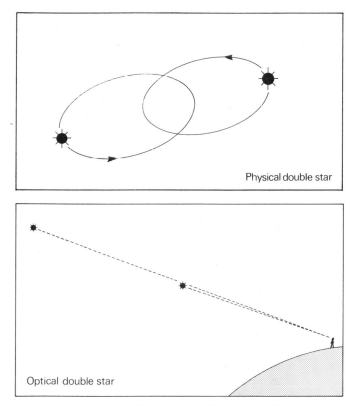

Physical double star

Optical double star

Most double stars are related (a physical double), but some lie by chance in the same line of sight (an optical double).

What are variable stars?

Not all stars are constant in brightness. Certain stars change noticeably in brightness from night to night, and some even from hour to hour. The brightest of the varying stars is Betelgeuse, which ranges a full magnitude (i.e. 2.5 times) between maximum and minimum intensity, taking many months or even years to go from one maximum to the next. The variations in the light output of Betelgeuse are caused by changes in the actual size of the star. Although the variations of Betelgeuse have no set period, a number of stars pulsate regularly, like a beating heart, every few days or weeks. These are known as *Cepheid variables*, and they follow the rule that the longer the cycle of variation, the more luminous is the star.

Some stars which appear to vary in brightness are actually close pairs in which one star periodically eclipses the other, temporarily blocking off its light from us. The prototype of these so-called *eclipsing binaries* is Algol, in the constellation Perseus, which drops to one-third its usual brightness every 2 days 20 hours 49 minutes. Other interesting variable stars are mentioned in the constellation notes.

The most spectacular of the varying stars are the *novae*, faint stars which can flare to thousands of times their usual brightness for a few days before sinking back into obscurity. Novae got their name, which is Latin for 'new', because in the past they were really thought to be new stars appearing in the sky. Now we know that a nova is actually an old, dim star that has flared up because gas has fallen onto it from a neighbour. Nova outbursts are unpredictable, and are often first spotted by amateur astronomers.

What is a shooting star?

During your sky watching, you will from time to time see a bright streak of light dash across the sky like a cosmic laser beam, lasting no more than a second or so. This is a *meteor*, popularly termed a shooting star. Do not misidentify shooting stars with satellites or high-flying aeroplanes, which look like moving stars but drift at a more leisurely pace.

Despite their name, shooting stars are nothing to do with stars at all. They are particles of dust shed by comets, usually no bigger than a grain of sand, plunging into the Earth's atmosphere at speeds from 10 to 75 km/sec (6 to 46 mile/sec). We see the glowing streak of hot gas as the dust particle burns up by friction with the atmosphere about 100 km (60 miles) above the Earth. The brightest meteors outshine the stars, and some break up into glowing fragments. Occasionally, much larger chunks of rock and metal penetrate the atmosphere and land on Earth. These are called meteorites.

On any clear night, several meteors are visible to the naked eye each hour, burning up at random. These are known as *sporadic* meteors. But at certain times of the year the Earth ploughs through dust storms moving along the orbits of comets. We then see a meteor shower, when as many as 100 meteors may be visible each hour.

Due to an effect of perspective, all the members of a meteor shower appear to diverge from a small area of sky known as the *radiant*. The meteor shower is named after the constellation in which its radiant lies: the Perseids appear to radiate from the constellation of Perseus, the Geminids from Gemini and so on. Meteor showers recur on the same dates each year, although the intensity of a shower can vary from year to year. As the Earth penetrates the swarm of dust, the number of meteors visible each hour builds up over a few days to its maximum, and then falls away again. Amateur astronomers monitor the progress of a shower by counting the number of meteors visible, and estimating their brightnesses.

When watching for meteors from a shower, do not look directly at the radiant, but look about 45 degrees to one side of it. A list of the year's main meteor showers is given in the table overleaf. Note that the maximum number of meteors shown in column three will be seen only in dark conditions, when the radiant is high in the sky. If the radiant is low, or if the sky is bright (for instance from moonlight), the number of meteors visible will be very much less.

METEOR SHOWERS

Name of shower	Date of maximum	Number of meteors visible per hour at maximum (approx.)
Quadrantids	January 3–4	100
Lyrids	April 21–22	15
Eta Aquarids	May 5–6	40
Delta Aquarids	July 28–29	20
Perseids	August 12	60
Orionids	October 21	20
Taurids	November 3	12
Leonids	November 17–18	10
Geminids	December 13–14	60

What is a comet?

People often confuse comets with shooting stars, but they are quite different things. Whereas a shooting star appears as a brief streak of light, a comet looks like a hazy smudge hanging in the sky. A comet's movement against the star background is noticeable only over a period of hours, or from night to night. Comets are snowballs of frozen gas and dust that move on highly elongated orbits around the Sun, usually taking hundreds or thousands of years to complete one circuit.

At least a dozen comets are seen by astronomers each year. About half of these are previously known comets returning to the inner part of the solar system, while the others are new discoveries. Most comets are too faint to be seen without a large telescope. Only rarely, perhaps every ten years or so, does one become bright enough to be prominent to the naked eye.

A bright comet is awesomely beautiful. From its glowing head stretches a transparent tail of gas and dust, extending for millions of kilometres. A comet's tail always points away from the Sun. The dust released by the comet disperses into space, some of it eventually being swept up by the Earth to appear as meteors.

What is the Milky Way?

On clear, dark nights, a hazy band of starlight arches across the heavens. The Greeks called it the Milky Way. We know that the Milky Way consists of innumerable stars comprising an enormous wheel-shaped structure, the Galaxy, of which our Sun is a member. The stars that are scattered randomly over the night sky, forming the constellations, are among the nearest to us in the Galaxy. The more distant stars are concentrated into the crowded band of the Milky Way. The Milky Way, therefore, is the rest of our Galaxy as seen from our position within it.

The Galaxy's centre lies in the direction of the constellation Sagittarius, where the Milky Way star fields are particularly dense. Our Sun lies approximately two-thirds of the way from the hub to the rim of the Galaxy, which is about 100,000 light years in diameter. Beyond the edge of the Galaxy is empty space, dotted with other galaxies.

On the monthly star maps, the Milky Way is indicated in light blue. Sparkling star fields spring into view if you sweep along this region with binoculars or small telescopes.

What is a star cluster?

In places, stars congregate in clusters, some of which are visible to the naked eye, most notably the group called the Pleiades in the constellation Taurus. There are two sorts of star cluster, distinguished by the types of stars in them and their location in the Galaxy. Open clusters are loose groupings of young stars dotted along the spiral arms of our Galaxy. Some open clusters, such as the Pleiades, are still surrounded by traces of the gas clouds from which they were born. Open clusters contain from a handful of stars up to a few thousand stars.

Altogether different in nature are the ball-shaped globular clusters, mostly found well away from the plane of the Milky Way. They are swarms of up to 300,000 stars, much more tightly bunched together than in open clusters. The stars in globular clusters are very old – indeed, they include some of the oldest stars known. Since globular clusters are much further away from us than open clusters, they appear fainter. The best globular cluster for northern observers is M13 in the constellation Hercules.

What is a nebula?

Between the stars are vast clouds of gas and dust known as nebulae, the Latin for mist, a word that accurately describes their foggy appearance. They are best seen in clear country skies, away from smoke and streetlights.

Some nebulae shine brightly, while others are dark. The most famous bright nebula is in Orion, visible to the naked eye as a softly glowing patch. The Orion Nebula is a gas cloud from which stars are forming, and the new-born stars at its centre light up the surrounding gas (see page 18). Other nebulae remain dark because no stars have yet formed inside them. Dark nebulae become visible when they are silhouetted against dense star fields or bright nebulae. The Milky Way star fields in Cygnus are bisected by a major dark nebula, the Cygnus rift.

Other nebulae are formed by the deaths of stars, including the so-called planetary nebulae, which despite their name have nothing to do with planets. The name arose because, when seen through small telescopes, they often look like the discs of planets. Actually, planetary nebulae are glowing shells of gas sloughed off by stars like the Sun at the ends of their lives. Stars that are much more massive than the Sun die in violent explosions, splashing fountains of luminous gas into space. The most famous remnant of an exploded star is the Crab Nebula in the constellation Taurus (see page 63).

What is a galaxy?

Some objects that at first sight resemble nebulae are actually distant systems of stars beyond our Milky Way – other galaxies, many millions of light years apart, dotted like islands throughout the Universe. The smallest galaxies consist of approximately a million stars, while the largest galaxies contain a million million stars or more. We live in a fair-sized galaxy of about 100,000 million stars.

Galaxies come in two main types: spiral and elliptical. Spiral galaxies, of which our Galaxy is one, have arms consisting of stars and gas winding outwards from their star-packed centres. A sub-species of spiral galaxies, called barred spirals, have a bar of stars across their centre; the spiral arms emerge from the

NAMING STAR CLUSTERS, NEBULAE AND GALAXIES

Star clusters, nebulae and galaxies are often referred to by numbers with the prefix M or NGC. The M refers to Charles Messier, an eighteenth-century French comet hunter who compiled a list of over 100 fuzzy-looking objects that might be mistaken for comets. Astronomers still use Messier's designations, and enthusiasts avidly track down the objects on his list. In 1888 a far more comprehensive listing of star clusters and nebulous objects was published, called the *New General Catalogue of Nebulae and Clusters of Stars*, or NGC for short. This was followed by two supplements known as the *Index Catalogues* (IC), bringing the total number of objects catalogued to 13,000, most of them only within reach of large telescopes. In this book we use the Messier numbers where they exist, or NGC and IC numbers otherwise.

ends of the bar. Elliptical galaxies have no spiral arms. They range in shape from almost spherical to flattened lens shapes. Time-exposure photographs with large telescopes are needed to bring out the full beauty of galaxies. Through binoculars and small telescopes, most galaxies appear only as hazy patches of light. Like nebulae, galaxies are best seen in clear, dark skies.

The nearest major galaxy to us is visible to the naked eye in the constellation of Andromeda. It appears as a faint smudgy patch, similar in apparent width to the full Moon. The Andromeda Galaxy is a spiral galaxy, 2 million light years from us (see page 53).

How to look at the stars

To begin stargazing you need nothing more than your own eyes, supplemented by a modest pair of binoculars. The monthly star charts in this book are designed to be used for stargazing with the naked eye. With these charts, you will be able to identify the stars and constellations visible on any night of the year.

Specific constellations are featured in more detail in the monthly star notes, and to study the objects described in them requires some form of optical aid. Optical instruments collect more light than the naked eye does, thus showing fainter objects as well as making objects appear bigger by magnifying them. In astronomy, the ability of an instrument to collect light is often more important than the amount by which it magnifies. This is particularly so in the case of binoculars, which are an indispensable starting instrument for any would-be stargazer.

Binoculars usually have relatively low magnifications of between six and ten times, so they will not show detail on the planets, but their light-grasp will bring into view many stars and nebulae that are beyond reach of the naked eye. Binoculars have a much wider field of view than telescopes, and so are better suited for studying extended objects such as scattered star clusters. Even if you subsequently obtain a telescope, binoculars will remain of use.

Binoculars bear markings such as 6×30, 8×40 or 10×50. The first figure is the magnification, and the second figure is the aperture of the front lenses in millimetres: the larger the aperture, the greater the light-grasp and hence the fainter the objects that you can see. Binoculars with high magnifications of 12 times or more are occasionally encountered, but these need to be mounted on a tripod to hold them steady.

Telescopes are rather like telephoto lenses, but whereas a telephoto lens on a camera is described by its focal length a telescope is described by its aperture. Remember that a telescope with an aperture of, say, 50 mm has the same light-grasp as a pair of 50 mm binoculars, but it can cost several times more. The main advantage of a telescope over binoculars is that it has higher magnification.

The smallest telescopes – those with apertures up to 75 mm (3 inches) – are of the refracting type, like a spyglass, which you look straight through. Larger telescopes, starting with apertures of 100 mm (4 inches), are usually reflectors, in which mirrors collect the light and bounce it into an eyepiece. Reflectors are cheaper to make in large sizes than refractors, and they have shorter tubes, which is more convenient.

A problem that becomes apparent through a telescope is the unsteadiness of the atmosphere. Turbulent air currents make the image of a star or planet seem to boil and jump around, which limits the amount of detail that can be discerned, particularly at higher magnifications. The steadiness of the atmosphere is known as the *seeing*, and it changes markedly from night to night.

Telescopes have interchangeable eyepieces which give a range of magnifications. High magnifications are useful when studying fine detail on the planets or separating the close components of a double star. But no matter how much magnification you use on a small telescope, it will not show as much detail as a larger telescope. In fact, too high a power on a telescope will make the image so faint that you will end up seeing less than if you had used a lower power. A good rule of thumb is to keep to a maximum magnification of 20 times per 10 mm of aperture (50 times per inch). Do not be tempted by small telescopes that offer magnifying powers of many hundreds of times. In practice, such high magnifications will be unusable.

Most of the objects mentioned in this book are within view of telescopes with an aperture of 100 mm or less, using low to medium magnifying powers.

How to use the maps in this book

There are twelve all-sky charts in this book, one for each month of the year, usable every year. They show the sky as it appears between latitudes 30 degrees north and 60 degrees north at fixed times of night that are convenient for observaton. To understand the maps, imagine the night sky as a dome overhead. The maps are a flattened representation of that dome. The map projections have been carefully chosen to minimize distortion of the constellation shapes on the printed page.

To use the maps, turn them so that the direction marked at the bottom of the map – north, south, east or west – matches the direction you are facing. The map then depicts the sky as it appears at 10 p.m. in mid-month. The sky will look the same at 11 p.m. at the start of the month and 9 p.m. at the end of the month. (NB: one hour must be added to these times when daylight saving is in operation.) When you look upwards you must think of the maps as being stuck to the dome of the sky, arching over you on all sides.

Four horizons are shown on the maps, for observers at latitude 30 degrees north, 40 degrees north, 50 degrees north and 60 degrees north. It should not be difficult to work out where the horizon falls on the maps for your specific latitude. You will be able to use the maps from anywhere between latitudes 30 degrees north and 60 degrees north, and up to 10 degrees outside these limits there will be hardly any discrepancy between what the maps show and the way the sky appears. Note

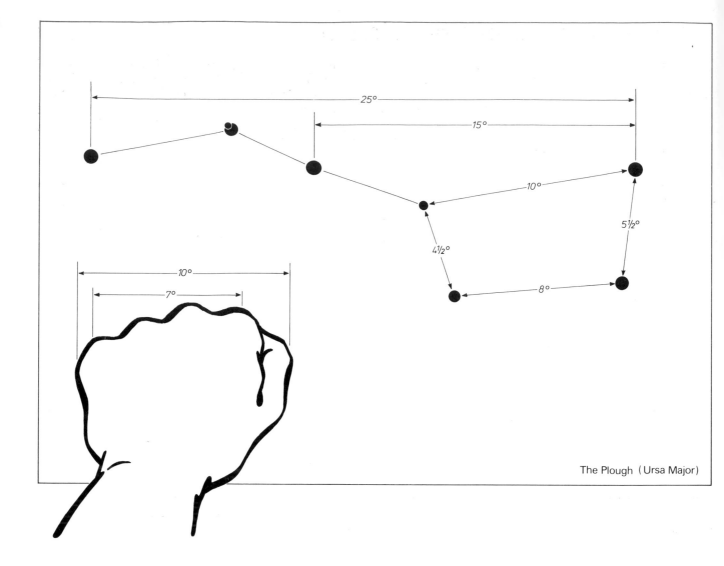

The Plough (Ursa Major)

how your latitude affects the stars that are visible.

The monthly maps show the brightest stars down to magnitude 5.0, realistically depicting the sky as seen by the naked eye on an averagely clear night. The maps of individual constellations show stars down to magnitude 6.5, plus fainter selected stars and numerous objects of interest for users of binoculars and small telescopes.

Scale in the sky

One difficulty when matching up the maps in a book with the appearance of the real sky is the matter of scale. Most people overestimate the size of objects in the sky. For example, it is a remarkable fact that the full Moon, which is half a degree across, can be covered by the width of a pencil held at arm's length. (Try it!) An outstretched palm will cover the main stars of Orion, and you can cover most of the stars in the Plough with your hand.

How can we judge whether a particular constellation covers a large or a small area of sky? The width of a fist held at arm's length makes an ideal distance indicator, about 7 degrees wide. As a guide to scale, each of the detailed constellation maps in this book carries an outline of a fist for comparison. If your hand is unusually big or small the comparison will not be exact, but it will still be a useful guide to the size of constellation you are looking for.

HOW THE SKY CHANGES WITH THE SEASONS

The sky is like a clock and a calendar. It changes in appearance with the time of night and with the seasons of the year. For example, look at these diagrams of the area around the north pole of the sky. At 10 p.m. at the start of the year, the familiar saucepan-shape of seven stars known variously as the Plough or Big Dipper is standing on its handle to the right of Polaris, the north pole star. At the same time of night three months later, the Plough is almost directly overhead, while the W-shaped constellation of Cassiopeia sits low on the northern horizon. By July, the Plough appears to the left of Polaris. At 10 p.m. in October the Plough is scraping the northern horizon, while Cassiopeia rides high. In another three months the stars are back to where they started and the yearly cycle begins anew.

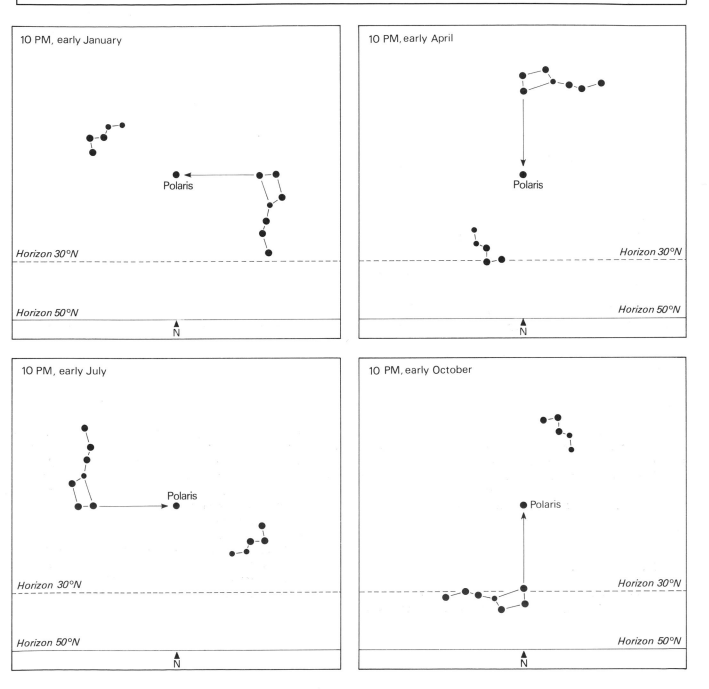

FINDING YOUR WAY

To find your way in unfamiliar territory, you need a map and signposts. This book provides the maps. The signposts are in the sky, once you know where to look. Start with an easily recognizable pattern, such as the Plough or Orion, and work your way outwards from it to locate other constellations and bright stars, a technique known as star-hopping. While star-hopping around the sky you will find that there are many natural 'pointers' that direct you from constellation to constellation. Also, bright stars often form distinctive patterns of lines, triangles and squares that you can remember. These pages demonstrate some of the best ways of locating prominent stars and constellations. As you navigate your way among the stars you will discover many more signposts of your own.

Signposts of spring

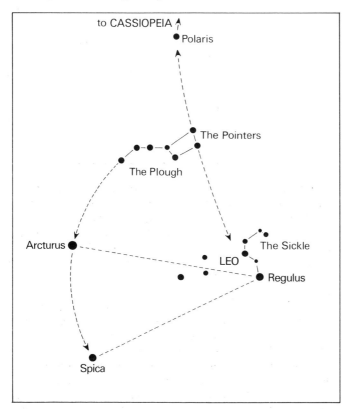

Start with the familiar saucepan-shape of the Plough or Big Dipper, which rides high in the sky on spring evenings. The seven stars of the saucepan are actually the most prominent members of the constellation Ursa Major, the Great Bear. From the diagram you can see that two stars of the saucepan's bowl – the ones that lie furthest from the handle – point towards the north pole star, Polaris. These two stars in the Plough are popularly known as The Pointers. If you extend the distance between them by about five times you will reach the pole star. Opposite Polaris from the Plough lie five stars that form a distinctive W-shape, which is the constellation of Cassiopeia.

Polaris is a star of moderate brightness lying in a somewhat blank region. It is not exactly at the north pole of the sky, but is situated about one degree from it. During the night the stars circle around the north celestial pole (and hence around Polaris) as the Earth spins on its axis.

Go back to the Plough. If you extend the Pointers in the opposite direction, away from Polaris, you will come to the constellation of Leo, the lion. This constellation is notable for the sickle-shape of stars, like a reversed question mark, that make up the lion's head.

Now look at the handle of the Plough. Follow the curve made by the stars of the handle until you come to Arcturus, one of the brightest stars in the sky. Continue the curve and you reach the sparkling star Spica, in the constellation of Virgo. Note that Arcturus and Spica form a prominent triangle with Regulus, the brightest star in Leo.

Signposts of summer

High in the sky on summer evenings lies an isosceles-shaped trio of bright stars known as the Summer Triangle. In order of decreasing brightness they are Vega, Altair and Deneb. Vega is the fifth-brightest star in the sky, and it is the first star to appear as the sky darkens in July and August, shining overhead like a blue-white diamond.

Deneb is the brightest star in Cygnus, a constellation that represents a swan but which is better visualized as a cross, as shown on the diagram here. Deneb marks the head of the cross.

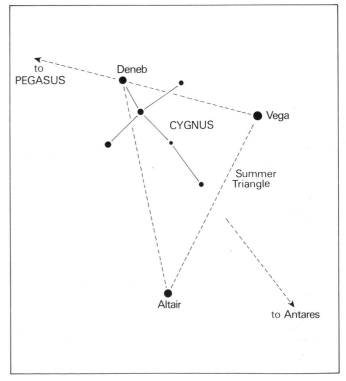

The foot of the cross bisects a line between Vega and Altair, and points towards the bright star Antares, which appears low on the horizon from mid-northern latitudes. Antares lies due south as the sky darkens in July. It has a prominent red colour, and represents the heart of the scorpion, Scorpius.

A line from Vega through Deneb directs you towards the Square of Pegasus, a quadrangle of four stars that is the centrepiece of the autumn sky. Apart from the Summer Triangle, the summer sky is remarkably bereft of prominent star patterns. Incidentally, despite its name the Summer Triangle remains visible well into the autumn.

Signposts of autumn

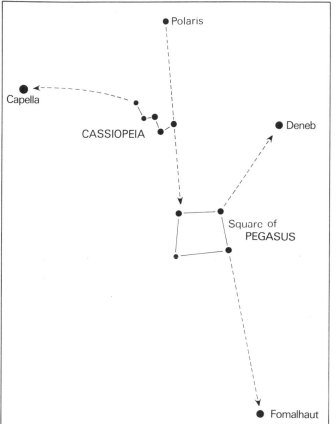

As the Summer Triangle sets in the west on autumn evenings, the Square of Pegasus takes centre stage. The Square lies high in the south at 10 p.m. in mid-October, 8 p.m. in mid-November, and 6 p.m. in mid-December. The four stars that mark the corners of the Square are of only moderate brightness. They enclose a large area of sky that is almost entirely devoid of naked-eye stars.

The Square of Pegasus is like a keystone in the autumn sky, from which many surrounding stars and constellations can be identified. To the top right of the Square lie Deneb and the rest of the Summer Triangle. Between the Square and the pole star lies the W-shaped constellation of Cassiopeia. As the diagram shows, a line extended upwards from the left-hand side of the Square passes through the end of the W shape and on to Polaris. Alternatively, you can use this line in reverse, running it from Polaris through Cassiopeia, to find the Great Square. A line extended downwards from the right-hand side of the Square directs you to Fomalhaut, a bright star in the southern constellation of Piscis Austrinus, but often difficult to find from mid-northern latitudes because it is so close to the horizon.

Now look at Cassiopeia, riding high above you. A line drawn across the top of the W points to Capella, one of the prominent stars of the winter sky.

Signposts of winter

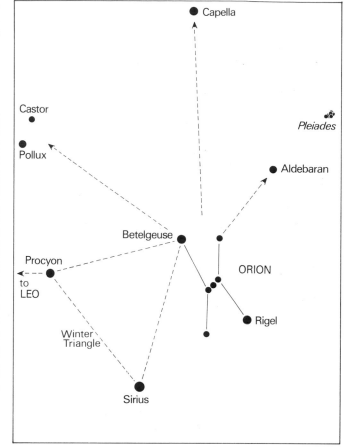

The sky in winter is more richly stocked with bright stars than in any other season of the year. The jewel in winter's crown is the brightest star in the entire sky, Sirius, which glitters due south at midnight at the beginning of January, at 10 p.m. at the beginning of February, and at 8 p.m. at the beginning of March. Sirius lies at the apex of the Winter Triangle of brilliant stars, which is completed by Procyon to its upper left, and Betelgeuse to its upper right.

Betelgeuse marks the top left of the constellation of Orion, a rectangular-shaped figure that is the easiest of the winter constellations to recognize. At the bottom right of the Orion rectangle is Rigel, a star slightly brighter than Betelgeuse. Two fainter stars complete the rectangle. Across the centre of Orion runs a distinctive line of three stars comprising Orion's belt.

To the top right of Orion is another prominent star, Aldebaran, which represents the glinting eye of Taurus the Bull. Aldebaran is of similar brightness to Betelgeuse, and both stars have a noticeably orange tinge. Continue the line from Orion through Aldebaran and you will come to a hazy-looking knot of stars called the Pleiades, a star cluster that shows up well in binoculars.

Above Orion, almost directly between it and the north celestial pole, lies the bright star Capella. To the top left of Orion, above the Winter Triangle, lies a famous pair of stars called Castor and Pollux, the celestial twins in the constellation Gemini. Off to the left of Castor, Pollux and Procyon lies Leo, the lion, which introduces us once again to the skies of spring.

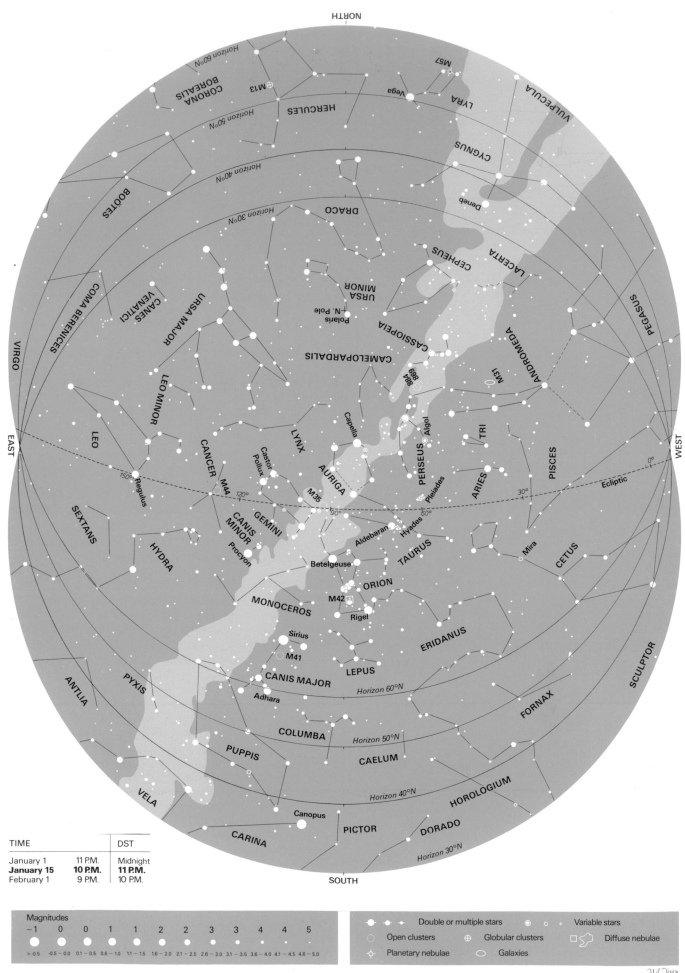

NORTH

Horizon 60°N
Horizon 50°N
Horizon 40°N
Horizon 30°N

CORONA BOREALIS
HERCULES
M13
VEGA
LYRA
M57
VULPECULA
CYGNUS
BOÖTES
Deneb
LACERTA
DRACO
ANDROMEDA
CEPHEUS
PEGASUS
URSA MINOR
M31
Polaris · N. Pole
CASSIOPEIA
CANES VENATICI
URSA MAJOR
CAMELOPARDALIS
884
869
PERSEUS
Algol
TRI
LEO MINOR
Capella
PISCES
LYNX
AURIGA
ARIES
30°
0°
Ecliptic
WEST
EAST
Castor
Pollux
Regulus
150°
M44
M35
120°
CANCER
Pleiades
Aldebaran
Hyades
Mira
CETUS
GEMINI
90°
60°
TAURUS
SEXTANS
CANIS MINOR
Betelgeuse
HYDRA
Procyon
ORION
M42
SCULPTOR
MONOCEROS
Rigel
ERIDANUS
Sirius
FORNAX
M41
LEPUS
PYXIS
CANIS MAJOR
Horizon 60°N
ANTLIA
Adhara
COLUMBA
Horizon 50°N
CAELUM
HOROLOGIUM
PUPPIS
VELA
Horizon 40°N
Canopus
PICTOR
DORADO
CARINA
Horizon 30°N

SOUTH

TIME		DST
January 1	11 P.M.	Midnight
January 15	**10 P.M.**	**11 P.M.**
February 1	9 P.M.	10 P.M.

Magnitudes

−1   0   0   1   1   2   2   3   3   4   4   5

> -0.5   -0.5 – 0.0   0.1 – 0.5   0.6 – 1.0   1.1 – 1.5   1.6 – 2.0   2.1 – 2.5   2.6 – 3.0   3.1 – 3.5   3.6 – 4.0   4.1 – 4.5   4.6 – 5.0

Double or multiple stars   Variable stars

Open clusters   Globular clusters   Diffuse nebulae

Planetary nebulae   Galaxies

Wil Tirion

JANUARY

The planets this month

Venus

1988 An evening object, moving from Capricornus into Aquarius.
1989 In the morning sky, moves from Ophiuchus into Sagittarius, becoming lost in the dawn twilight towards the end of the month.
1990 Vanishes into the evening twilight in the first week of January, reappearing in the morning sky at the end of the month.
1991 In the evening sky in Sagittarius at the start of the month, then moves into Capricornus, and ends the month in Aquarius.
1992 A prominent morning object, moving via Scorpius and Ophiuchus into Sagittarius during the month.

Mars

1988 Starts the month in Libra, crossing Scorpius into Ophiuchus by the end of the month. It passes north of the bright red star Antares ('the rival of Mars'). Mars is the fainter of the two.
1989 Moves from Pisces into Aries during the month, fading rapidly as it recedes from Earth.
1990 Moves from Ophiuchus into Sagittarius.
1991 In Taurus, near the Pleiades. Mars fades rapidly as it recedes from Earth, but is still mag. 0 at the end of the month.
1992 Emerges into the morning sky early in the month as a second-mag. object in Ophiuchus, moving into Sagittarius.

Jupiter

1988 In Pisces at mag. –2.
1989 Brighter than mag. –2 in Taurus, south of the Pleiades.
1990 Prominent in Gemini, brighter than mag. –2.
1991 Brighter than mag. –2 in Cancer. At opposition (i.e. due south at midnight) on the night of January 28–29.
1992 In Leo at mag. –2.

Saturn

1988 Moves into the morning sky at the start of the month, becoming visible at mag. 0.7 on the Ophiuchus-Sagittarius border.
1989 In Sagittarius. Too close to the Sun for observation at the start of the month, but moves into the morning sky by mid-month.
1990 In Sagittarius. Too close to the Sun at the start of the month, emerging into the dawn sky by month's end.
1991 Too close to the Sun throughout the month.

1992 In Capricornus at mag. 0.8. Moves into the evening twilight by mid-month and becomes lost.

January meteors

The year's most abundant meteor shower, the Quadrantids, is visible on January 3 and 4. Its peak of activity is much sharper than that of most showers, lasting only a few hours; the exact date and time of maximum varies from year to year. At best over 100 Quadrantids an hour can be seen, though the meteors of this shower are not as bright as other great displays such as the Perseids and Geminids. The Quadrantids radiate from the northern part of Boötes, near the handle of the Plough. This area of sky was once occupied by the now-defunct constellation Quadrans Muralis, the Mural Quadrant, from which the shower takes its name. Unfortunately this area of sky is not well placed for observation until after midnight, so Quadrantid watchers must be prepared to brave the small hours until dawn to see the shower at its best. The Quadrantids move on an orbit around the Sun with a period of about five years. Their parent comet is thought to have disintegrated about 1500 years ago, leaving the Quadrantids as the only one of the main meteor showers without a known parent body.

Orion

Orion is the most glorious of all constellations. It glitters with bright stars and is replete with objects of interest for observers with all forms of optical equipment, including the humblest binoculars. It is an ideal place to begin a yearly tour of the heavens.

Orion is popularly personified as a hunter, giant or warrior. On old star charts he is depicted brandishing his club and shield against the snorting charge of neighbouring Taurus the bull. In Greek mythology Orion was the son of Poseidon. He was stung to death by a scorpion in retribution for his boastfulness. At the request of his lover, Artemis, he was placed in the sky so that he sets in the west as his slayer, in the form of the constellation Scorpius, rises in the east. In another story, Orion was smitten by the beauty of the seven nymphs known as the Pleiades, whom he began to pursue. To save the nymphs, Artemis turned them into a cluster of stars. The Pleiades star cluster lies in the neighbouring constellation of Taurus, and Orion still seems to chase them in his nightly course across the sky.

Orion is one of the easiest constellations to recognize, lying due south at 10 p.m. this month. Look first of all at Alpha (α) Orionis, better known as Betelgeuse, at the top left of the constellation. Note that it has a distinctly orange colour, which is brought out more strongly through binoculars. Betelgeuse marks the right shoulder of Orion. Its name is often mis-translated as Armpit of the Central One, but in fact its original Arabic title meant Hand of Orion.

Betelgeuse is the type of star known as a red supergiant. It is so large that it fluctuates in size between about 300 and 400 times the diameter of the Sun. This leads to unpredictable changes in its brightness between about magnitude 0 and 1.5, although normally it hovers around magnitude 0.5. Keep an eye on the brightness changes of Betelgeuse by comparing it with other stars, particularly Aldebaran and Capella, from time to time. Various distances are given for Betelgeuse, but a more recent determination places it just over 300 light years away, closer than previously thought.

Compare Betelgeuse with the blue-white star Rigel at the bottom right of the constellation. Rigel marks Orion's left leg, and not surprisingly its name comes from the Arabic meaning left leg. Rigel, also known as Beta (β) Orionis, is the brightest star in Orion, magnitude 0.1. It, too, is a supergiant, but it has a much hotter surface, which accounts for its difference in colour from Betelgeuse. The temperature of Rigel's surface is 12,000 kelvin, while that of Betelgeuse is a relatively cool 3000 K. Rigel is about 900 light years away, three times farther than Betelgeuse.

Rigel is worth looking at carefully for another reason: it is a double star. It has a companion star of magnitude 6.8 that is difficult to see in the smallest telescopes because of the glare from Rigel itself. The aperture needed to pick out this companion will depend on the steadiness of the atmosphere and how high Rigel is above the horizon. Probably at least 75 mm aperture is required, but the only way to be sure is to go out and see for yourself.

One distinctive feature of Orion is the line of three bright stars that marks its belt. Each of these stars is of second magnitude. Of the three stars of the belt, Epsilon (ε) Orionis (Alnilam) and Zeta (ζ) Orionis (Alnitak) are at a similar distance from us, about 1200 light years, whereas Delta (δ) Orionis (Mintaka) seems about twice as far away. Delta Orionis has a wide 7th-magnitude companion star that is easily seen in small telescopes. At the left end of the belt, Zeta Orionis is a double star that presents a challenge for amateur telescopes. A 4th-magnitude companion lies 2.3 seconds of arc from it; normally, stars this far apart should just be divisible in an aperture of 50 mm, but the difference in brightness between the stars means that at least 75 mm aperture is needed to separate them, as well as a night of steady seeing.

A strip of faint nebulosity, known as IC 434, extends southwards from Zeta Orionis. On one side of this is the celebrated Horsehead Nebula, formed by a cloud of dark dust that is silhouetted against the glowing hydrogen gas behind it. Unfortunately, the Horsehead is virtually impossible to see with even large telescopes, but it shows up well on long-exposure photographs, looking like a celestial chess piece.

A survey of Orion's stellar treasures would be incomplete without a visit to Sigma (σ) Orionis, lying just below the leftmost star of the belt. To the naked eye Sigma Orionis appears as an unremarkable star of 4th magnitude, but small telescopes reveal it to be flanked by three fainter stars, looking like a planet with moons. Also visible in the same field of view is a triple star, Struve 761, which consists of a narrow triangle of 8th- and 9th-magnitude stars. This rich telescopic sight is one of Orion's unexpected delights.

One reason for Orion's prominence is that most of its stars lie in a nearby spiral arm of the Galaxy where new stars are still being born. The centre of starbirth in this region is the Orion Nebula, 1300 light years away, which appears as a misty patch to the naked eye; it marks the sword of Orion, hanging from his belt (see below).

Colour photographs show the Orion Nebula to be a mixture of yellow and red, whereas to the human eye it appears greenish. The reason lies in the different colour sensitivities of the eye and colour film. The greenish colour seen by the eye comes from atoms of ionized oxygen, whereas colour film is better at picking up the reddish emission from hydrogen gas of which the Orion Nebula mostly consists, as do all similar nebulae. Colour photographs give a better idea of the true colour of the Nebula than does the human eye.

The Orion Nebula is about 20 light years in diameter and contains enough gas to make thousands of stars. If we could turn the clock back 5000 million years to the birth of the Sun we would find that our region of space looked like the Orion Nebula.

THE ORION NEBULA

The Orion Nebula is a luminous cloud of gas, the finest object of its kind in the heavens, also known as M42 and NGC 1976. Wreaths of ghostly glowing gas spread over a degree of sky, twice the apparent diameter of the full Moon, making a sight not to be missed in any instrument. To its north is a smaller patch of gas, M43, alias NGC 1982, actually part of the same cloud. Because the Nebula is so large, binoculars are excellent for viewing it. Embedded at the Nebula's centre is a 5th-magnitude star, Theta-1 (θ¹) Orionis. This star was born recently from the surrounding gas, and it illuminates the Nebula. Home in on Theta-1 Orionis with small telescopes and you will see why it is popularly known as the Trapezium: it consists of four stars of magnitudes 5.1, 6.7, 6.7 and 8.0, arranged in a rectangle. The Trapezium is almost at the tip of a dark wedge in the Nebula known as the Fish Mouth. To the lower left of the Trapezium, binoculars show Theta-2 (θ²) Orionis, a wide double star of magnitudes 5.2 and 6.5.

This whole area is a complex of nebulosity and young, hot stars, which together comprise the sword of Orion. At the tip of the sword, on the southern edge of the Orion Nebula, is Iota (ι) Orionis, the brightest star in the sword. Iota Orionis is a hot, blue-white giant of 3rd magnitude with a 7th-magnitude companion shown by small telescopes. To the lower right of Iota lies Struve 747, an easy pair of 5th- and 6th-magnitude white stars for small telescopes. Above the Orion Nebula lies a smaller smudge of nebulosity, NGC

1977, containing the 5th-magnitude star 42 Orionis. (Its apparent neighbour, 45 Orionis, is an unrelated foreground object.) Farther north binoculars reveal NGC 1981, a scattered handful of stars of 6th magnitude and fainter, marking the top of the sword handle.

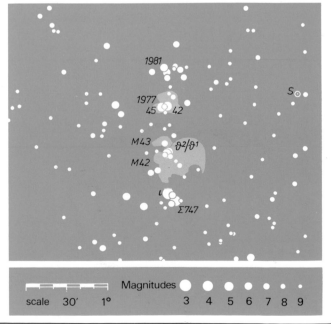

scale 30' 1° Magnitudes 3 4 5 6 7 8 9

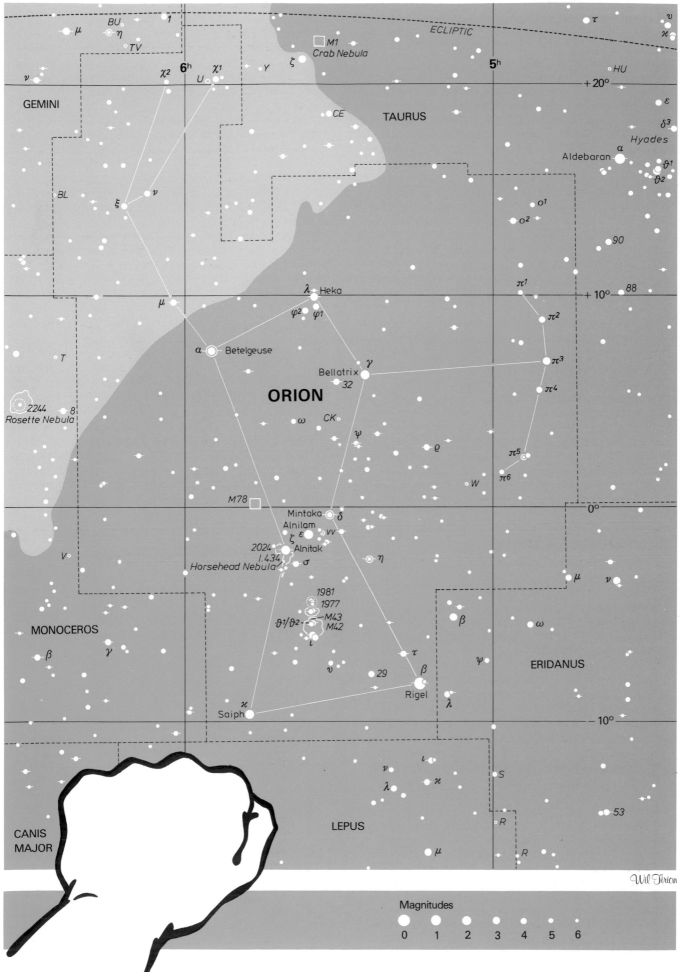

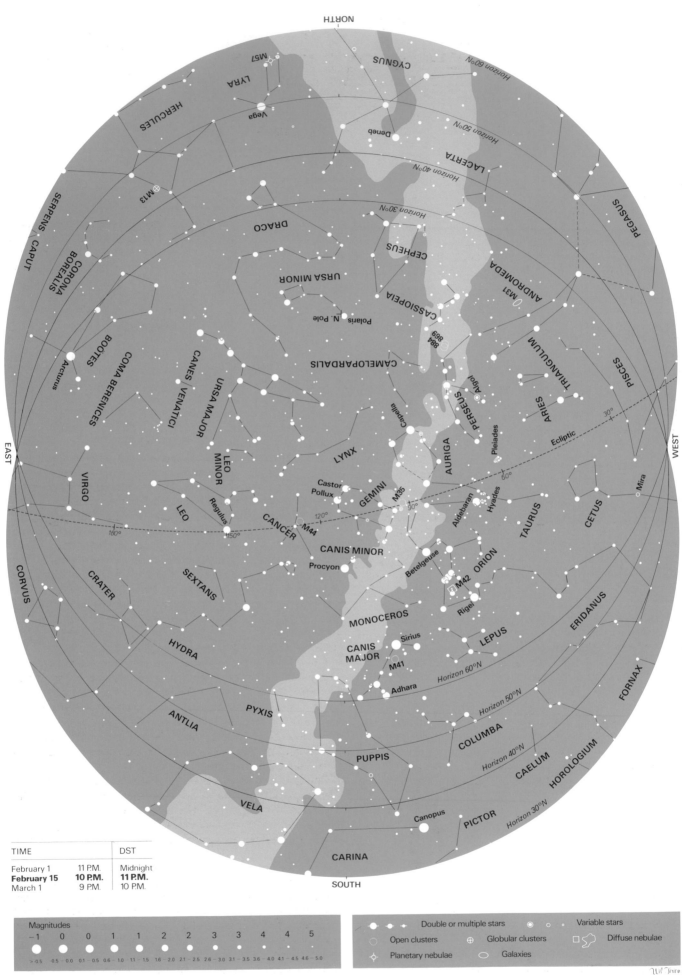

TIME		DST
February 1	11 P.M.	Midnight
February 15	**10 P.M.**	**11 P.M.**
March 1	9 P.M.	10 P.M.

Magnitudes

−1	0	0	1	1	2	2	3	3	4	4	5
>-0.5	-0.5 - 0.0	0.1 - 0.5	0.6 - 1.0	1.1 - 1.5	1.6 - 2.0	2.1 - 2.5	2.6 - 3.0	3.1 - 3.5	3.6 - 4.0	4.1 - 4.5	4.6 - 5.0

Double or multiple stars　　　Variable stars

Open clusters　　　Globular clusters　　　Diffuse nebulae

Planetary nebulae　　　Galaxies

FEBRUARY

KEY STARS

The Winter Triangle formed by Sirius, Procyon and Betelgeuse remains prominent in the evening sky this month. Other stars well-placed for observation are Rigel, in Orion, and Aldebaran, in Taurus, both in the south-west. Castor and Pollux, the twin stars of Gemini, are high in the sky at 10 p.m. Regulus, the brightest star in Leo, is high in the south-east by 10 p.m. At the same time, orange Arcturus is rising on the eastern horizon.

The planets this month

Venus

1988 In the evening sky, starting the month in Aquarius, then moving into Pisces.
1989 Lost in the morning twilight throughout February.
1990 A brilliant morning object (up to mag. –4.3) in Sagittarius.
1991 Moves from Aquarius into Pisces, in the evening sky.
1992 A morning object, moving from Sagittarius into Capricornus. Passes one degree north of the much fainter Mars on February 20 and is very close to Saturn on February 29.

Mars

1988 Moves from Ophiuchus into Sagittarius, brightening from mag. 1.5 to mag. 1.2. Passes the brighter planet Saturn (mag. 0.8) towards the end of the month.
1989 A first-magnitude object in Aries, ending the month on the border with Taurus.
1990 In Sagittarius, around mag. 1.5. Mars is close to the brighter planet Saturn (mag. 0.8) at the end of February.
1991 In Taurus, north of the Hyades cluster and Aldebaran. Mars fades markedly during this month (from mag. 0.1 to mag. 0.7) as it recedes from Earth, but remains brighter than Aldebaran.
1992 Moves from Sagittarius into Capricornus at mag. 1.5. Brilliant Venus passes one degree north on February 20.

Jupiter

1988 In Pisces, ending the month on the border with Aries.
1989 Prominent at mag. –2 in Taurus, south of the Pleiades.
1990 Mag. –2 or brighter throughout the month. In Gemini, near the borders with Orion and Taurus.
1991 In Cancer at mag. –2, passing in front of Praesepe (the Beehive Cluster).
1992 Bright (mag. –2) in Leo. At opposition (due south at midnight) on February 29, 660 million km from Earth.

Saturn

1988 In Sagittarius at mag. 0.8.
1989 In Sagittarius at mag. 0.8.
1990 In Sagittarius. Saturn (mag. 0.8) forms a close pair with the fainter planet Mars (mag. 1.5) at the end of the month.
1991 Emerges from twilight into the morning sky in the first week of February, in Capricornus.
1992 In Capricornus. Moves into the morning sky in mid-month at first mag. Brilliant Venus passes close on February 29.

Canis Major

In the south this month sits the constellation Canis Major, the Greater Dog, home of Sirius, the brightest star in the entire sky. Canis Major and nearby Canis Minor repesent two dogs following at the heels of Orion. Surprisingly, the only object of significance in the Lesser Dog, Canis Minor, is its brightest star, Procyon; hence Canis Minor might be termed the lone-star constellation. By contrast, Canis Major is packed with bright stars. On old maps, the dog is depicted as standing on his hind legs, with flaming Sirius (popularly known as the Dog Star) marking his snout.

Sirius has a magnitude of –1.46. Its name comes from the Greek meaning searing or scorching, for in ancient times it was actually thought to be a source of heat. The sweltering 'dog days' of high summer were attributed to the Dog Star, for they occurred when Sirius lay close to the Sun. Hesiod, one of the earliest Greek poets, wrote of 'heads and limbs drained dry by Sirius'. Because of the outstanding brilliance of Sirius, its yearly passage around the sky was used as a calendar-marker from at least the time of the ancient Egyptians, over 2000 years BC.

Sirius appears twice as bright as the second most prominent star, Canopus, which can be seen below it at this time of year if you are south of latitude 37 degrees north. The exceptional brilliance of Sirius is due to a combination of its light output and its relative closeness to us. Sirius is roughly twice the mass and twice the diameter of the Sun, and gives out more than 20 times as much light. That by itself is not exceptional. What helps make Sirius so bright in our skies is that it is 8.7 light years away, the fifth-closest star to the Sun. Hence it outshines more powerful stars, such as Betelgeuse in Orion, which are much more distant. Of the stars visible to the naked eye, only Alpha Centauri in the southern hemisphere is closer to us than Sirius.

Sirius is almost pure white, but it twinkles a multitude of colours as its light is broken up by air currents in the Earth's atmosphere on frosty winter nights. In binoculars or a small telescope it is dazzling. Sirius has a bizarre 8th-magnitude companion, a small and dense star of the type known as a *white dwarf* (see box). This tiny companion of the Dog Star has been nicknamed The Pup. Unfortunately, it is so close to Sirius that a large telescope is needed to show it.

Beta (β) Canis Majoris is a blue giant star 720 light years away, magnitude 2.0. It bears the Arabic name Mirzam, meaning the announcer, from the fact that its rising heralds the appearance of Sirius. The second-brightest star in the constellation, Epsilon (ε) Canis Majoris, magnitude 1.5, is another blue giant, 490 light years away.

The real superstar of the constellation, though, is Delta (δ) Canis Majoris, which the Arabs named Wezen, meaning weight. The legend behind this odd name has been lost but it is certainly appropriate, for the star is indeed weighty, with an estimated mass at least 20 times that of the Sun. It is a brilliant supergiant, giving out more than 100,000 times as much light as the Sun, so that it appears of magnitude 1.9 despite its considerable distance of 3000 light years, nearly ten times as far as Betelgeuse.

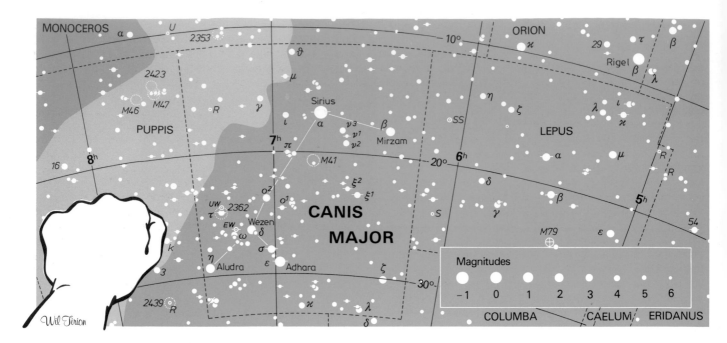

Almost as impressive is Eta (η) Canis Majoris, another supergiant, 50,000 times more luminous than the Sun. It appears of magnitude 2.4 and is 2500 light years away. If Sirius were removed to such a distance it would appear of 11th magnitude, way below naked-eye visibility. Conversely, if Delta and Eta Canis Majoris were as close to us as Sirius they would each shine as brilliantly as a half Moon, so the night sky would never be truly dark when they were above the horizon.

Canis Major contains two attractive star clusters for small instruments. Just south of Sirius is M41, covering the same area of sky as the full Moon. Under good conditions it can even be seen as a fuzzy spot by the naked eye. M41 contains about 50 stars, the

brightest of them a 7th-magnitude orange giant. Binoculars pick out the brightest members, and many more are visible in small telescopes. Low magnification is needed to fit all the cluster into the field of view. Note how the stars are arranged in chains, a common effect in clusters. Observers in high northern latitudes will find that the proximity of M41 to the horizon dims its splendour.

Much more compact than M41 is the cluster NGC 2362, centred on the 4th-magnitude blue-white giant Tau (τ) Canis Majoris. NGC 2362 contains about 40 members, many of them visible in small telescopes. Tau Canis Majoris itself has a light output of 20,000 Suns. Star and cluster are 5000 light years away.

WHITE DWARFS

In 1862 the American astronomer Alvan G. Clark first saw the faint companion of Sirius while testing a new 47-cm telescope. But this companion star, Sirius B, was a puzzle. Its surface was hotter than that of the Sun, yet the star itself was much dimmer than the Sun. That meant it must be very small – about 2 per cent of the Sun's diameter, only twice the size of the Earth, which is astoundingly small by stellar standards. Such a tiny, hot star is termed a white dwarf.

Yet the puzzle did not end there. Sirius B was found to contain as much mass as the Sun. That much mass packed into such a small ball meant that Sirius B must be incredibly dense, far denser than any substance known on Earth. In fact, a spoonful of material from Sirius B would weigh many tonnes.

By coincidence, the brightest star in Canis Minor, Procyon, is also accompanied by a white dwarf, fainter and even more difficult to see than the companion of Sirius. White dwarfs are now known to be the dying remnants of stars like the Sun. The companions of Sirius and Procyon have both seen better days.

Sirius B orbits Sirius A once every 50 years (see diagram). The two stars are always too close for Sirius B to be seen in small telescopes. Even when the two are at their widest separation, around the year 2025 AD, a telescope of at least 200 mm aperture and steady air will be needed to pick out the faint spark of Sirius B from the glare of its brilliant neighbour.

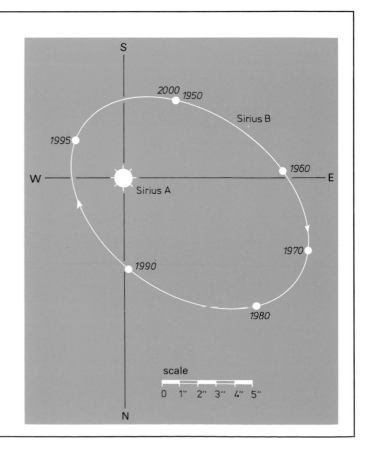

Monoceros

Between the stars of the Winter Triangle lies the faint but fascinating constellation of Monoceros, the Unicorn, introduced on star maps in the seventeenth century. Often overlooked in favour of its brilliant neighbours, Monoceros none the less contains some classic objects.

Among them is perhaps the finest triple star for small telescopes, Beta (β) Monocerotis. It appears double under low magnification, but higher power divides the fainter companion. The three stars, two of 5th magnitude and one of 6th magnitude, form an arc. Epsilon (ε) Monocerotis is a double star easily separated by small telescopes, consisting of blue-white and yellow components of magnitudes 4.3 and 6.7, set among an attractive scattering of fainter stars.

Monoceros contains two celebrated star clusters, both enveloped by faint nebulosity from which they have recently formed. NGC 2244 is a widespread group of a dozen or so stars, 6th magnitude and fainter, covering an area similar to the apparent width of the Moon and best seen in binoculars or wide-field telescopes with low magnification. The hidden treasure of this object is the surrounding gas cloud, known as the Rosette Nebula, beautiful on long-exposure photographs. Unfortunately it is too faint to be seen through small telescopes, although under excellent conditions it might be glimpsed through large binoculars as a faint halo, more than two Moon diameters wide.

About two degrees from NGC 2244 lies 6th-magnitude Plaskett's Star, named after the Canadian astronomer John Stanley Plaskett who found in 1922 that it is the heaviest pair of stars known. According to latest measurements, each star is at least 55 times as massive as the Sun, which places them both firmly in the super-heavyweight class. The pair are so close together that they cannot be seen individually in any telescope. Only by analysing light from Plaskett's Star through a spectroscope can astronomers tell that two stars, not one, are present, and that they orbit each other every 14 days. Stars of such immense mass do not live for long, so Plaskett's Star must be very young, probably less than a million years old.

Another of the secret treasures of Monoceros is the star cluster NGC 2264. Small telescopes and binoculars show about a dozen stars arranged in the shape of an arrowhead, with numerous fainter members. Brightest of the cluster is S Monocerotis, an intensely hot blue-white star of magnitude 4.7 but slightly variable. NGC 2264 is enveloped in a glowing gas cloud, too faint to see in small telescopes but well known from long-exposure photographs. Into its southern end intrudes a cone-shaped wedge of dark dust which gives the object its popular name, the Cone Nebula. NGC 2264 is one of the youngest known clusters, estimated to be only one or two million years old.

NGC 2264 and the Cone Nebula lie about 2500 light years away, a distance similar to that of NGC 2244 and the Rosette Nebula. Clearly, considerable star formation is under way in that part of the Galaxy.

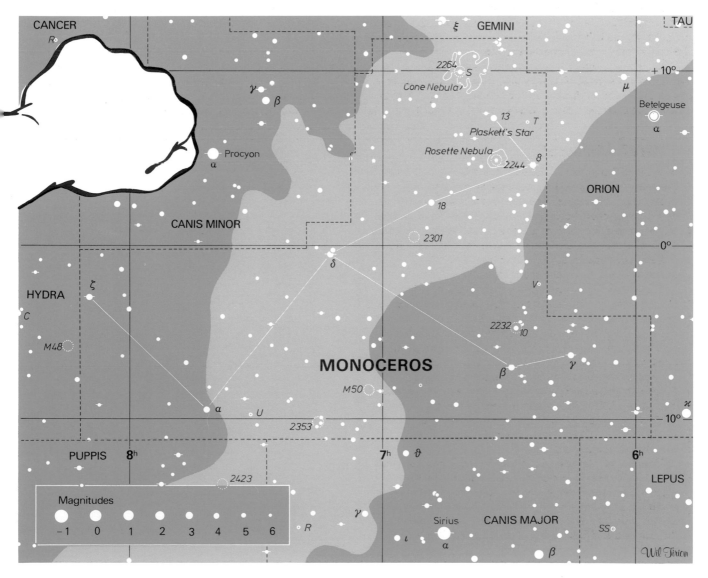

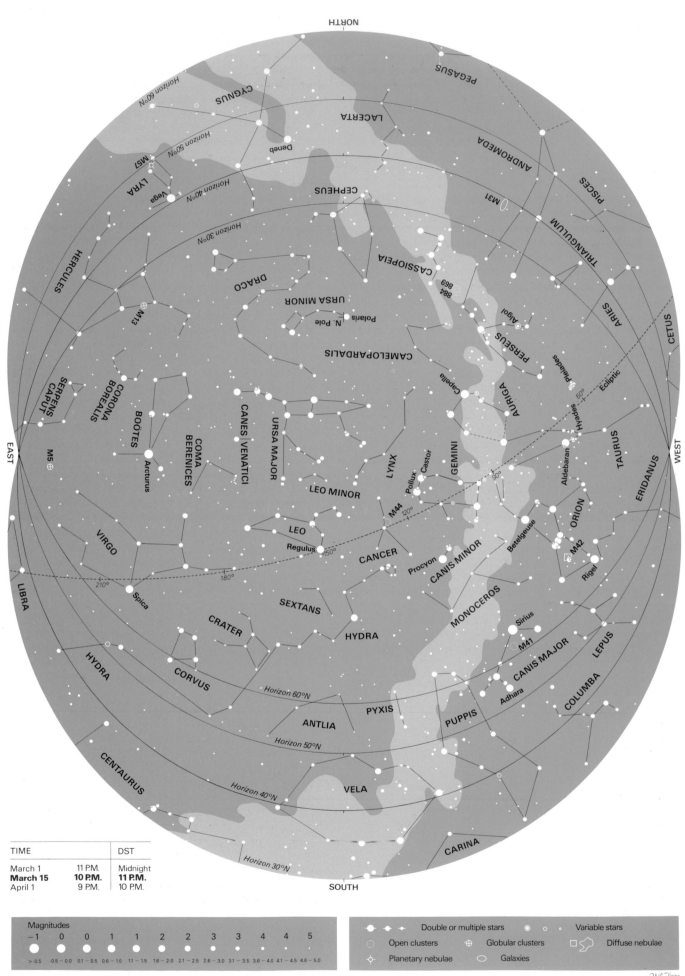

NORTH

TIME		DST
March 1	11 P.M.	Midnight
March 15	**10 P.M.**	**11 P.M.**
April 1	9 P.M.	10 P.M.

Magnitudes

−1	0	0	1	1	2	2	3	3	4	4	5
> -0.5	-0.5 - 0.0	0.1 - 0.5	0.6 - 1.0	1.1 - 1.5	1.6 - 2.0	2.1 - 2.5	2.6 - 3.0	3.1 - 3.5	3.6 - 4.0	4.1 - 4.5	4.6 - 5.0

Double or multiple stars   Variable stars

Open clusters   Globular clusters   Diffuse nebulae

Planetary nebulae   Galaxies

Wil Tirion

MARCH

KEY STARS

Taurus, Orion and the Winter Triangle are sinking towards the western horizon by 10 p.m. this month. Castor and Pollux are still well displayed, and Capella glints high in the northeast. Regulus, brightest star of Leo, shines high in the south around 10 p.m. Below it is the straggling figure of Hydra, a region bereft of bright stars. The seven stars that make up the familiar shape of the Plough, or Big Dipper, are well placed for viewing this month. On the eastern horizon, Spica and Arcturus are rising, heralding the approach of spring.

The planets this month

Venus

1988 An evening object, moving from Pisces, through Aries, and into Taurus. Venus (mag. –3.8) passes Jupiter (mag. –1.7) in the early part of March.

1989 Too close to the Sun for observation throughout the month.

1990 Brilliant at mag. –4.2 in the morning sky, moving from Sagittarius into Capricornus. Greatest elongation (i.e. maximum angular separation from the Sun) is on March 30.

1991 In the evening sky, moving from Pisces into Aries.

1992 A morning object, moving from Capricornus via Aquarius to end the month on the border with Pisces.

Mars

1988 A first-magnitude object in Sagittarius.

1989 Passes between the Pleiades and Hyades clusters in Taurus, overtaking the much brighter planet Jupiter (mag. –1.8) as it does so. Mars fades to mag. 1.5 during the month.

1990 Moves from Sagittarius into Capricornus, brightening slightly from mag. 1.4 to mag. 1.2.

1991 In Taurus, moving towards the border with Gemini, and fading to mag. 1.2 as it recedes from Earth.

1992 Moves from Capricornus into Aquarius, mag. 1.4. Passes brighter Saturn on March 7.

Jupiter

1988 Moves from Pisces into Aries, at mag. –1.7. The much brighter planet Venus (mag. –3.8) overtakes it in the first week of March.

1989 In Taurus, between the Pleiades and Hyades star clusters, at mag. –1.8.

1990 In Gemini, near the border with Taurus, at mag. –1.8.

1991 In Cancer at mag. –2, near Praesepe (the Beehive Cluster).

1992 In Leo at mag. –2.

Saturn

1988 In Sagittarius at mag. 0.7.

1989 In Sagittarius at mag. 0.8.

1990 In Sagittarius at mag. 0.8.

1991 In Capricornus at mag. 0.9.

1992 In Capricornus at mag. 1.0. Fainter Mars passes on March 7.

Gemini

Gemini represents the twins of Greek mythology, Castor and Pollux, after whom the constellation's two brightest stars are named. Old star maps show the twins holding hands in the sky. The stars Castor and Pollux mark the heads of the twins. In legend, their mother was Queen Leda of Sparta, but each twin had a different father. Castor was the son of Leda's husband, King Tyndareus, while Pollux was the son of the god Zeus, who visited Leda in the form of a swan. Castor and Pollux were members of the crew of Argonauts who sailed in search of the Golden Fleece. After their death, Zeus set them in the sky. Seafarers regarded the heavenly twins as patron saints, and called upon them for protection in times of danger. The electrical phenomenon known as St. Elmo's fire, seen among the masts and rigging of ships, was thought to be the guiding spirit of the twins.

Castor and Pollux are contrasting stars. Although they appear in the same region of sky, the two stars are not related. Pollux is the closer to us, at a distance of 36 light years, while Castor lies 45 light years away. Pollux, also known as Beta (β) Geminorum, is an orange giant of magnitude 1.1, slightly the brighter of the pair. Castor, Alpha (α) Geminorum, is actually an astounding family of six stars, all linked by gravity. To the naked eye it appears as a single star of magnitude 1.6. But small telescopes, with high magnification, divide Castor into a dazzling pair of blue-white stars of magnitudes 1.9 and 2.9. These orbit each other every 500 years or so. Castor was, in fact, the first pair of stars recognized to be orbiting each other, a discovery announced by William Herschel in 1803. Also present, but less easy to see in small telescopes, is a 9th-magnitude red dwarf companion some distance from the brighter pair of stars.

All three stars are themselves close doubles, making Castor a stellar sextuplet. Each pair is so close together that the stars cannot be seen separately through even the largest telescopes. Astronomers discovered that the stars were double by analysing their light through a spectroscope, when the presence of light from two stars was recognized. Hence the stars are known as *spectroscopic binaries*.

Castor A, through telescopes the brightest of the components, consists of two stars bigger and brighter than the Sun orbiting each other every nine days. Castor B, the second-brightest component, consists of another two stars larger than the Sun, orbiting each other every three days. The third component, Castor C, consists of two red dwarfs, smaller and fainter than the Sun, which orbit each other in less than a day. They also eclipse each other, causing the brightness of Castor C to vary from magnitude 9.2 to 9.6. All the members of the Castor family were born together from the same cloud of gas, and have remained linked by gravity ever since. Try to imagine the view from a planet around one of those stars as you gaze at Castor through a telescope.

Gemini is the home of a large and impressive star cluster, M35, near the feet of the twins, close to the border with Taurus. M35 contains over 100 member stars scattered across an area as large as the full Moon. M35 is a fine sight in binoculars,

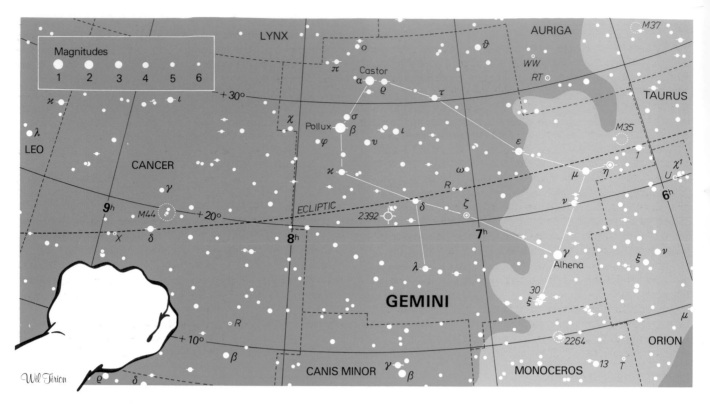

appearing as a mottled patch enveloped in mist. Small telescopes clearly show individual stars, of 8th magnitude and fainter. Unlike many clusters, M35 is not condensed at the centre. Rather, users of telescopes will note that its stars are arranged in disconnected chains, somewhat reminiscent of the lights on a Christmas tree. M35 lies 2800 light years away.

More challenging is the planetary nebula NGC 2392. Small telescopes show it as a fuzzy, bluish disk like an 8th-magnitude star out of focus, of similar size to the globe of the planet Saturn.

Here, we are seeing a shell of gas thrown off by a dying star. That star appears of 10th magnitude at the centre of the nebula. Through large telescopes, NGC 2392 looks like a face surrounded by a fringe, which has given rise to its two popular names, the Eskimo Nebula and the Clown Face Nebula. Although you will not see such detail through a small telescope, NGC 2392 is one of the easier planetary nebulae to spot, and is well worth searching for.

THE BEEHIVE CLUSTER

Between Gemini and Leo lies the constellation of Cancer, the Crab. It is faint, with no star brighter than magnitude 3.5, but at its centre lies a classic object for binoculars, the star cluster M44. This cluster was known in ancient times as Praesepe, the Manger, but it is now popularly termed the Beehive. The cluster is flanked by two stars, known as the Asses, visualized as feeding at the Manger. To the north is Gamma (γ) Cancri, magnitude 4.7, called Asellus Borealis, the Northern Ass, while Delta (δ) Cancri, magnitude 3.9, is the Southern Ass, Asellus Australis.

M44 is visible to the naked eye on clear nights as a misty patch, and was familiar to the ancient Greeks, who kept an eye on it for the purpose of weather forecasting. If the Manger was invisible on an otherwise clear night, this was said to be a sign of a forthcoming storm.

M44 covers one and a half degrees of sky, three times the apparent size of the full Moon – too large to fit into the field of view of a normal telescope but ideal for binoculars. The Beehive Cluster is larger even than the Pleiades in Taurus, although its stars are not so bright. The brightest member of the Beehive is Epsilon (ε) Cancri, a white star of magnitude 6.3. In all, M44 contains about 20 stars brighter than 8th magnitude, and several dozen more that are brighter than 10th magnitude. Binoculars show it as a breathtaking sight that resembles a swarm of bees around a hive. This stellar beehive is 520 light years away.

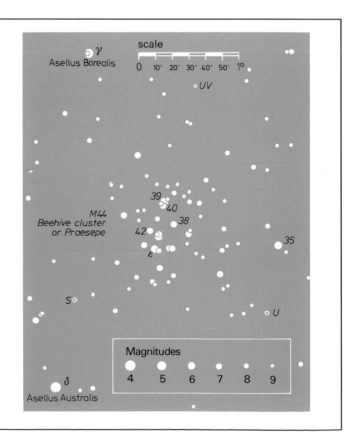

Leo

One of the few constellations that genuinely resembles its name is Leo, the lion. Legend identifies it with the lion slain by Hercules as one of his twelve labours. The lion's head is represented by a sickle-shape of six stars, like a back-to-front question mark. The body is outlined by four stars, the tail being marked by Beta (β) Leonis, a white star of magnitude 2.1 whose name, Denebola, comes from the Arabic meaning lion's tail.

At the bottom of the sickle is the brightest star in Leo, Alpha (α) Leonis, better known as Regulus, meaning little king, appropriate for the king of beasts. Regulus appears in our skies as magnitude 1.4. It is a blue-white star 85 light years away, with an estimated diameter five times that of the Sun, and it gives out 150 times as much light as the Sun. Small telescopes or even binoculars show that Regulus has a wide companion star of 8th magnitude. Although 700,000 million km from Regulus, more than 100 times the distance of Pluto from the Sun, this companion is genuinely related, and moves through space together with Regulus.

The real showpiece of this constellation is the second-brightest star in the sickle, Gamma (γ) Leonis, also known as Algieba, meaning lion's mane. To the naked eye Gamma Leonis shines at magnitude 1.9. Binoculars show a 5th-magnitude star nearby, 40 Leonis, which is not related but which lies by chance in the same line of sight as Gamma Leonis. Turn a small telescope on to Gamma Leonis, and switch to an eyepiece that magnifies about 100 times. It splits into a pair of golden yellow stars, one of the most magnificent doubles in the sky. Both components of Gamma Leonis are orange giants, magnitudes 2.3 and 3.5. They orbit their common centre of mass every 600 years or so. Gamma Leonis is 170 light years away.

Leo contains a number of galaxies, though none are prominent in small instruments. Between Theta (θ) and Iota (ι) Leonis lie the brightest of the Leo galaxies, M65 and M66, both of 9th magnitude. They are spiral galaxies, but M65 is tilted sideways to us and so appears elliptical. In small instruments, M65 and M66 appear as misty patches of light.

If you succeed in finding these, try for the more difficult M95 and M96 under the body of Leo. M95 is a barred spiral of 10th magnitude, whereas M96 is an ordinary spiral of 9th magnitude. The structure of each galaxy will not be noticeable in small instruments, for only the brightest central region of each galaxy is visible. All four galaxies lie about 20 million light years away.

As with all such faint, hazy objects their visibility depends critically on viewing conditions. Under poor skies they will be invisible in even moderate-sized telescopes, but under the best conditions they may be glimpsed in binoculars.

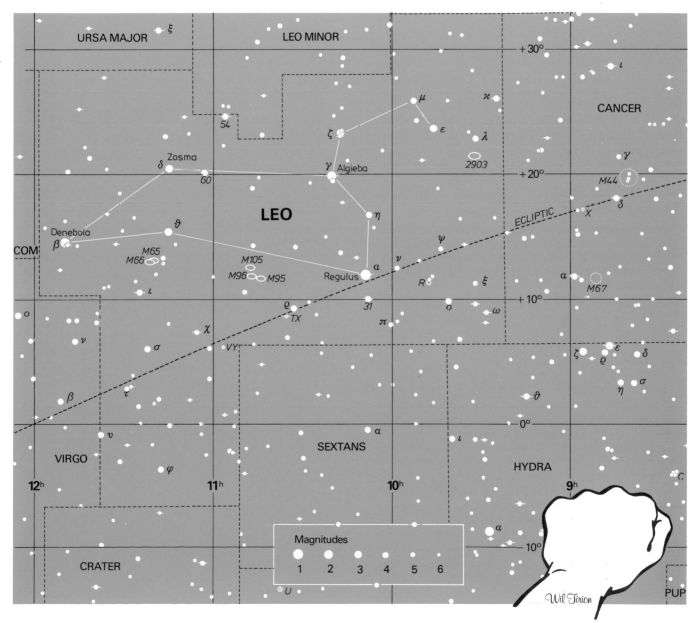

TIME		DST
April 1	11 P.M.	Midnight
April 15	**10 P.M.**	**11 P.M.**
May 1	9 P.M.	10 P.M.

Magnitudes

−1	0	0	1	1	2	2	3	3	4	4	5
>−0.5	−0.5–0.0	0.1–0.5	0.6–1.0	1.1–1.5	1.6–2.0	2.1–2.5	2.6–3.0	3.1–3.5	3.6–4.0	4.1–4.5	4.6–5.0

Double or multiple stars   Variable stars

Open clusters   Globular clusters   Diffuse nebulae

Planetary nebulae   Galaxies

Wil Tirion

APRIL

The planets this month

Venus

1988 A brilliant evening object in Taurus, brightening to mag. –4.2. At greatest elongation (i.e. maximum angular separation from the Sun) on April 3.

1989 Too close to the Sun for observation throughout the month.

1990 A morning object, starting the month in Capricornus at mag. –4.0, crossing Aquarius and ending in Pisces.

1991 An evening object, moving from Aries into Taurus at mag. –3.5, passing between the Hyades and Pleiades clusters.

1992 In Pisces, becoming lost in the morning twilight in the second half of the month.

Mars

1988 Moves from Sagittarius into Capricornus, brightening to mag. 0.5 by the end of the month.

1989 In Taurus, crossing the border into Gemini at the end of the month.

1990 Moving from Capricornus into Aquarius, mag. 1.0.

1991 In Gemini, fading during the month to mag. 1.6, the same brightness as Castor.

1992 Moves through Aquarius at mag. 1.3, ending the month in Pisces.

Jupiter

1988 In Aries at mag. –1.6, becoming lost in the evening twilight by mid-month.

1989 In Taurus, north of the Hyades cluster. At mag. –1.6, Jupiter is twice as bright as Aldebaran.

1990 In Gemini at mag. –1.6.

1991 In Cancer at mag. –1.8, near the Beehive Cluster (Praesepe).

1992 Almost stationary throughout the month in Leo, at mag. –1.9.

Saturn

1988 In Sagittarius at mag. 0.6.

1989 In Sagittarius at mag. 0.7.

1990 In Sagittarius at mag. 0.8.

1991 In Capricornus at mag. 0.9.

1992 In Capricornus at mag. 1.0.

April meteors

One of the year's lesser meteor showers, the Lyrids, makes its appearance this month. At best about a dozen meteors are visible each hour, radiating from a point on the border between Lyra and Hercules, near the bright star Vega. Although not numerous, the Lyrids are impressive. A typical Lyrid is bright, fast and leaves a luminous train. Maximum occurs on April 21 or 22 – the exact date varies from year to year – with activity declining rapidly a day or so either side of maximum. The Lyrids are one of the oldest known meteor showers. Recorded sightings go back more than 2000 years, when they were much more abundant. Their parent body is Comet Thatcher, which has the longest orbital period of any comet associated with a meteor shower, 415 years.

Ursa Major

The two bears, Ursa Major and Ursa Minor, stand high in the sky on spring evenings. Ursa Major's seven brightest stars comprise one of the most easily recognized patterns in the sky, popularly called the Plough or Big Dipper. Myths and legends concerning this star group stretch back to the earliest recorded times. We know the constellation as the Great Bear, although it is a strange-looking bear, with a long tail. In Greek mythology, Ursa Major represents the nymph Callisto, who was seduced by Zeus and was subsequently set among the stars in the form of a bear. One version of the tale identifies the Little Bear as her son, Arcas. Another story from ancient Greece says that they are both she-bears, representing the nymphs Adrasteia and Io who raised Zeus when he was an infant.

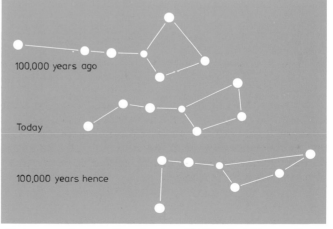

100,000 years ago

Today

100,000 years hence

All stars are on the move. Here, the stars of the Plough are shown as they appeared 100,000 years ago, as they are today, and as they will appear 100,000 years hence.

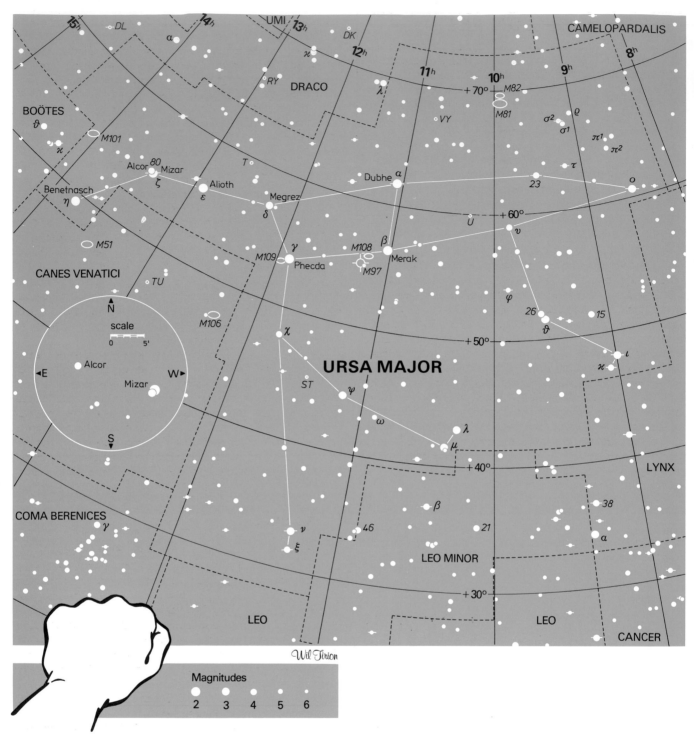

Ursa Major is the third-largest constellation in the sky. The seven stars that make up the familiar saucepan shape of the Plough or Dipper are only part of the complete constellation, forming the rump and tail of the bear. All the stars of the saucepan are of second magnitude except Delta (δ) Ursae Majoris, which is third magnitude.

Unlike most constellations, in which the brightest stars have no connection with each other, five of the seven stars in the saucepan are related, moving at the same speed and in the same direction through space. These five stars were evidently born together, and now form a very scattered cluster. The two non-members of the cluster are Benetnasch, at the end of the handle, and Dubhe, in the bowl. Over long periods of time, the motions of Benetnasch and Dubhe relative to the other five stars considerably after the familiar saucepan shape (see diagram on previous page).

The most celebrated star of Ursa Major is situated at the bend of the saucepan's handle, second from the end. It is Zeta (ζ) Ursae Majoris, popularly known as Mizar, magnitude 2.3. Keen eyesight shows a 4th-magnitude star, Alcor, nearby. Mizar and Alcor are sometimes termed the horse and rider. Mizar is about 60 light years from Earth, and Alcor is estimated to be 20 light years further away, so the stars are too widely separated to form a genuine binary, although they are moving through space along with the other members of the Ursa Major cluster.

But if you look at Mizar through a small telescope you will see that it also has a much closer companion of 4th magnitude that is definitely associated with it. Precise observations over more than two centuries have confirmed that this star is very slowly moving around Mizar on an orbit that takes about 10,000 years to complete.

THE WHIRLPOOL GALAXY

Just over the border from Ursa Major, in the constellation of Canes Venatici, the Hunting Dogs, lies M51, a celebrated spiral galaxy popularly known as the Whirlpool. It was the first galaxy in which spiral structure was noted, by Lord Rosse in 1845. A remarkable feature of the Whirlpool is the satellite galaxy at the end of one of its spiral arms, as pictured in long-exposure photographs. M51, of 8th magnitude, can be glimpsed under good conditions in binoculars as a misty patch. Small telescopes should show the starlike central nuclei of the Whirlpool and its satellite, but large telescopes are needed to trace its spiral arms. The Whirlpool Galaxy lies 18 million light years from us.

This close companion to Mizar was discovered in 1650 by the Italian astronomer Giovanni Riccioli. It made Mizar the first double star to be discovered with a telescope, and it was one of several 'firsts' for Mizar. Mizar was also the first double star to be photographed, by George P. Bond at Harvard in 1867; it was the first star discovered to be a spectroscopic binary, by Edward C. Pickering at Harvard in 1889; and it is often the first double star that amateur astronomers turn to with their telescopes. A spectroscopic binary is a pair of stars that are too close together to be seen separately in a telescope, and whose double nature can be inferred only from its spectrum. Actually, Mizar's companion is also a spectroscopic binary, making Mizar a four-star family. In addition, Alcor turns out to be another spectroscopic binary, completing a remarkable stellar grouping.

Another famous double, although far more challenging, is Xi (ξ) Ursae Majoris. It was the first double star to have its orbit calculated, in 1828, a demonstration that the same laws of gravity that govern the orbits of the planets around the Sun also apply to the stars. The two components, of magnitudes 4.3 and 4.8, orbit each other every 60 years, and their movement is currently so rapid that it can be followed from year to year in amateur telescopes. When the stars are closest together, between 1991 and 1995, a telescope of 150 mm aperture is needed to separate them, but before and after these dates they are separable in 100 mm telescopes with high magnification. After 1998, only 75 mm aperture is needed to divide them as they move apart. This is a good star on which to test your telescope.

Ursa Major is the home of several galaxies that lie far outside our Milky Way. In the northern part of the constellation lie M81 and M82, a prominent duo. M81 is one of the most beautiful spiral galaxies and, at 7th magnitude, one of the easiest to see in small instruments. It appears as a milky-white blur almost half the apparent diameter of the full Moon, much brighter towards its centre, and somewhat elliptical in shape because it is tilted at an angle to us. Under clear, dark skies you can pick it out in binoculars. M81 is similar to the great spiral galaxy in Andromeda, M31, although not as large.

Half a degree to the north of M81, and hence visible in the same low-power field of view, is an altogether more peculiar galaxy, M82, which appears cigar-shaped. M82 is smaller than M81 and is only one-quarter as bright. For many years M82 was thought to be an exploding galaxy, but more recent research has concluded that it is actually a spiral galaxy ploughing through a cloud of dust. It appears elongated because we see it edge-on. M82 and M81 are related, but estimates of their distance from us vary from about 7 million light years up to 18 million light years.

Ursa Minor

The Little Bear, Ursa Minor, looks indeed like a smaller but fainter version of Ursa Major. Its stars form a saucepan shape reminiscent of the Big Dipper, but with the curve of its handle reversed. Unlike Ursa Major, the stars of the Little Dipper are not related. Two stars in the Little Dipper's bowl, Beta (β) Ursae Minoris, Kochab, of second magnitude, and Gamma (γ) Ursae Minoris, Pherkad, of third magnitude, are known as the Guardians of the Pole. Keen eyesight, or binoculars, shows that Pherkad has a wide 5th-magnitude companion, which is actually an unrelated background star. Note also that the faintest star in the bowl, Eta (η) Ursae Minoris, consists of two widely spaced 5th-magnitude stars, also unrelated.

The most famous star in Ursa Minor is Alpha (α) Ursae Minoris, better known as Polaris, the Pole Star. It lies less than a degree from the celestial north pole, so it is directly overhead to an observer at the Earth's north pole. Although it shines modestly at magnitude 2.0 as seen from Earth, were we to travel to its vicinity we would find that it is in reality a most imposing star. Polaris is a yellow supergiant giving out as much light as 5000 Suns. Its distance is not accurately determined, but it must lie about 650 light years away.

Telescopes reveal that Polaris is a double star with a 9th-magnitude companion, although the companion is difficult to detect in the smallest apertures because it is so much fainter than Polaris. The companion slowly orbits Polaris over many thousands of years. Polaris is also a variable star of the Cepheid type, pulsating in size every four days. Its resulting variations in brightness are slight, about a tenth of a magnitude, virtually undetectable to the eye.

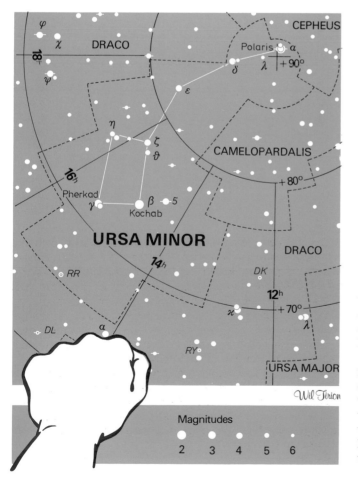

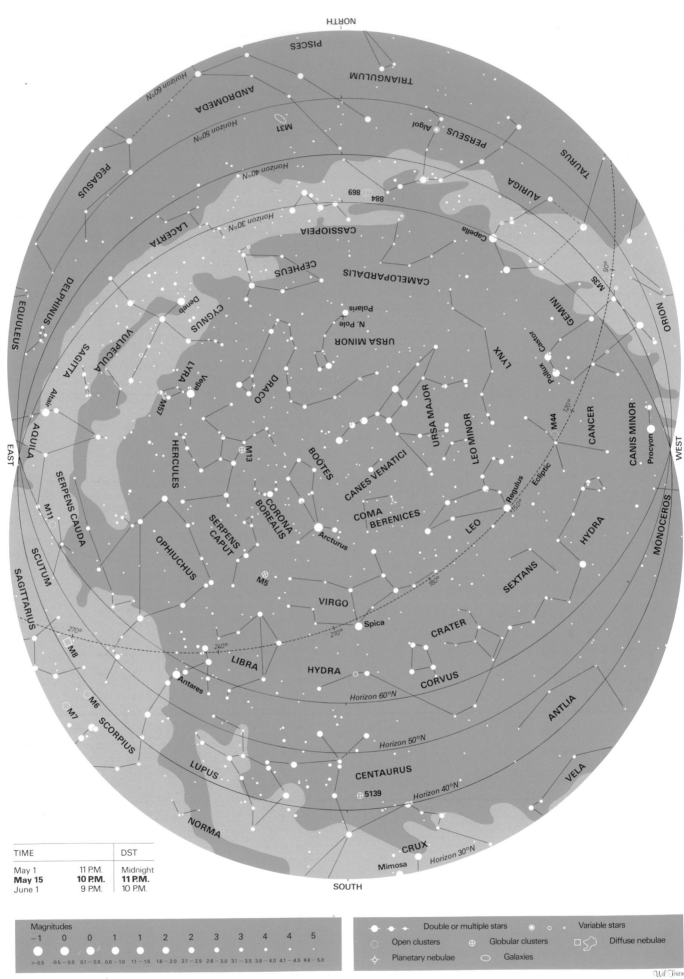

NORTH

PISCES
TRIANGULUM
ANDROMEDA
Horizon 60°N
M31
PERSEUS
Algol
TAURUS
PEGASUS
Horizon 50°N
LACERTA
AURIGA
869
Horizon 40°N
884
CASSIOPEIA
Capella
Horizon 30°N
CEPHEUS
CAMELOPARDALIS
90°
GEMINI
M35
DELPHINUS
Polaris
N. Pole
Castor
ORION
EQUULEUS
Denebola
CYGNUS
URSA MINOR
Pollux
Deneb
DRACO
LYNX
GEMINI
SAGITTA
VULPECULA
LYRA
Vega
M44
CANCER
120°
SAGITTA
Altair
M57
BOÖTES
URSA MAJOR
LEO MINOR
CANIS MINOR
AQUILA
HERCULES
M13
CANES VENATICI
Procyon
WEST
EAST
Regulus
Ecliptic
M11
SERPENS CAUDA
CORONA
BOREALIS
COMA
BERENICES
150°
HYDRA
SCUTUM
OPHIUCHUS
SERPENS
CAPUT
LEO
MONOCEROS
SAGITTARIUS
Arcturus
SEXTANS
180°
M5
VIRGO
CRATER
270°
M8
Spica
210°
HYDRA
240°
LIBRA
CORVUS
ANTLIA
Antares
M6
Horizon 60°N
M7
SCORPIUS
CENTAURUS
Horizon 50°N
VELA
LUPUS
5139
Horizon 40°N
NORMA
CRUX
Horizon 30°N
Mimosa
SOUTH

TIME		DST
May 1	11 P.M.	Midnight
May 15	**10 P.M.**	**11 P.M.**
June 1	9 P.M.	10 P.M.

Magnitudes

−1	0	0	1	1	2	2	3	3	4	4	5
>−0.5	−0.5 – 0.0	0.1 – 0.5	0.6 – 1.0	1.1 – 1.5	1.6 – 2.0	2.1 – 2.5	2.6 – 3.0	3.1 – 3.5	3.6 – 4.0	4.1 – 4.5	4.6 – 5.0

Double or multiple stars ● Variable stars
Open clusters ⊕ Globular clusters ▢ Diffuse nebulae
✦ Planetary nebulae ⬭ Galaxies

Wil Tirion

MAY

The planets this month

Venus

1988 An evening object in Taurus, near the borders with Auriga and Gemini. Brilliant (mag. –4.2) at the start of the month, but fades into the twilight at the end of the month.

1989 Too close to the Sun for most of the month, just becoming visible in evening twilight at the month's end.

1990 A morning object in Pisces at mag. –3.6, moving into Aries at the end of the month.

1991 In the evening sky. Moves from Taurus into Pisces, ending the month at mag. –3.8, four degrees south of Pollux.

1992 Lost in morning twilight throughout the month.

Mars

1988 Moves from Capricornus into Aquarius, brightening to mag. 0 by the end of the month as it approaches the Earth.

1989 In Gemini, a second-magnitude object fainter than Castor and Pollux.

1990 Moves from Aquarius into Pisces, brightening to mag. 0.7 by the end of the month.

1991 Starts the month in Gemini at mag. 1.6, the same as Castor. Ends the month in Cancer at mag. 1.8.

1992 In Pisces at mag. 1.2.

Jupiter

1988 At conjunction (directly behind the Sun) on May 2. Emerges into the morning sky at the end of May.

1989 In Taurus, north of the Hyades cluster. Becomes lost in the evening twilight in the second half of the month.

1990 In Gemini at mag. –1.5.

1991 In Cancer at mag. –1.6, adjacent to Praesepe (the Beehive Cluster).

1992 Remains almost stationary throughout the month in Leo at mag. –1.7.

Saturn

1988 In Sagittarius at mag. 0.4.

1989 In Sagittarius at mag. 0.5.

1990 In Sagittarius at mag. 0.6.

1991 In Capricornus at mag. 0.8.

1992 In Capricornus at mag. 0.9.

May meteors

Dust from Halley's Comet can be seen burning up in the atmosphere this month, producing the Eta (η) Aquarid meteor shower. At the shower's best, around May 5 each year, about 20 swift-moving meteors per hour radiate from near the star Eta (η) Aquarii, in the northern part of Aquarius. The shower's peak is broad, and activity remains high for several days either side of maximum. But the shower is difficult to see from the northern hemisphere, for two reasons. Since the radiant lies virtually on the celestial equator, observers need to be south of latitude 40 degrees north to have much hope of seeing any Eta Aquarids at all. Additionally, the radiant does not rise until nearly 2 a.m., and from northern latitudes does not get very high in the sky before dawn, which considerably restricts observations. In all, this is a difficult shower to observe.

Boötes

High in the spring sky stands the large constellation of Boötes, the Herdsman. It is easily found, lying next to the tail of Ursa Major. In legend, Boötes and Ursa Major are inextricably linked. In some tales Boötes is visualized as a man ploughing or driving a wagon (Ursa Major was often visualized as a wagon as well as a plough), while in other legends he is a herdsman or hunter chasing the Great Bear around the pole.

The constellation's brightest star is Arcturus, a name which is Greek for bear keeper. It was an important star to the ancient Greeks, who used its rising and setting as a guide to the changing seasons. Nowadays, the reappearance of Arcturus in the early evening is a welcome sign of approaching spring.

At magnitude –0.04, Arcturus is the fourth-brightest star in the entire sky, marginally brighter than Vega. Look carefully and you will see that it has a noticeably orange tint, which is more prominent when viewed through binoculars. Arcturus is the type of star known as a red giant. It is red (or more strictly orange) because it has a relatively cool surface temperature, two-thirds that of the Sun. And it is indeed a giant, with a diameter 27 times that of the Sun. Arcturus gives out over 100 times as much light as the Sun, and it lies 36 light years away.

Although Arcturus is so much bigger and brighter than the Sun, it actually has a similar mass to the Sun. So why are the two stars so different? The reason is that the Sun is in middle age, while Arcturus is nearing the end of its life. As stars age, they swell up in size to become red giants. One day our Sun will turn into a red giant, though fortunately not for several thousand million years yet. So when you look at Arcturus, reflect that you are seeing a preview of the Sun as it will be in 5000 million years' time. When the Sun becomes a red giant, the Earth and all life on it will be roasted to a cinder. Possibly Arcturus once had a life-bearing planet.

Boötes is well stocked with attractive double stars, but the most celebrated of them is very difficult to divide in small amateur telescopes because the components are so close together. The star concerned is Epsilon (ε) Boötis, sometimes known as Izar, the girdle, or Pulcherrima, meaning most beautiful because of the contrasting colours of its stars. To the naked eye it appears of magnitude 2.4, but actually it consists of an orange giant of magnitude 2.5 and a greenish-blue companion of magnitude 4.9. They are 2.8 seconds of arc apart, which would be within the power of a 50 mm refractor to split if they were of similar brightness. However, in practice the brightness difference means that the fainter star is difficult to separate from the primary's glare with even a 75 mm telescope. High magnification and very steady air are needed to divide the two, but the pair provide a stunning sight to those who succeed.

As an easier task, turn to Kappa (κ) Boötis in the north of the constellation, consisting of 5th- and 7th-magnitude white stars easily divisible in small telescopes. A similar but closer pair is Pi (π) Boötis, in the south of the constellation. For a warmer-toned pair, find Xi (ξ) Boötis, a 5th-magnitude golden-yellow star with a 7th-magnitude orange companion that orbits it every 150 years. Their distinct hues make them a showpiece duo.

More complex is the magnitude 4.3 white star Mu (μ) Boötis, whose name Alkalurops comes from the Arabic referring to the herdsman's crook or staff. Binoculars show that it has a 7th-magnitude companion. But this companion is itself a close double, whose components orbit each other every 260 years. These components can be split in telescopes as small as 60 mm, with high magnification.

Virgo

In the south around 10 p.m. this month lies the second-largest of all the constellations, Virgo, the Virgin. She is usually depicted as a maiden holding an ear of wheat, represented by the bright star Spica. As such she is identified with Persephone, the goddess of spring and the daughter of the harvest goddess Demeter. But Virgo has a second identity, that of Astraea, goddess of justice. In this guise she is depicted holding the scales of justice, represented by the neighbouring constellation of Libra.

Despite its large size Virgo is not a particularly prominent constellation, with the notable exception of the blue-white star Spica, magnitude 1.0. Spica is actually twice as hot as Sirius and 90 times more luminous. But in our skies it appears fainter than Sirius because of its considerably greater distance, 260 light years.

Virgo contains one of the most celebrated double stars, Gamma (γ) Virginis, also called Porrima after a Roman goddess of prophecy. To the naked eye Gamma Virginis appears as a star of magnitude 2.7, but it actually consists of a pair of identical white stars, each of magnitude 3.5, that orbit each other every 171 years. Their orbital motion is noticeable in small telescopes. At present, the two stars are moving closer together, making them progressively more difficult to split. The predicted separation of the two stars in 1990 is 3 seconds of arc, still within the capacity of a 50 mm telescope. By 1995 they will have closed to 2.5 seconds of arc, which will need a 60 mm telescope to split. By the year 2000 they will be 1.9 seconds of arc apart, requiring an aperture of at least 75 mm. The two stars will be closest together in 2007, by which time they will be indivisible in all but the largest telescopes. After that they will gradually move apart again, becoming an easy object for small telescopes once more.

Virgo's secret treasure is a cluster of distant galaxies, which spills across its northern border into Coma Berenices. This cluster contains about 3000 galaxies, centred on the giant elliptical galaxy M87, which is visible as a 9th-magnitude smudge in small telescopes or even binoculars. M87 is known to radio astronomers as the powerful radio source Virgo A. Long-

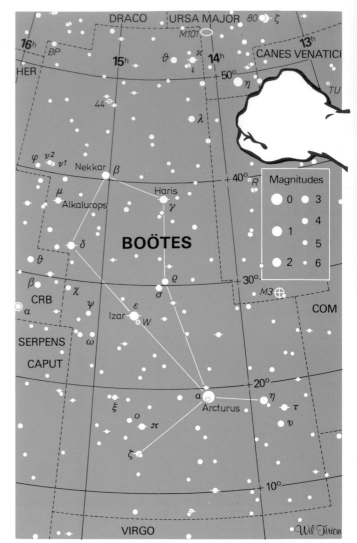

exposure photographs show a jet of luminous gas being shot out of M87, as though the galaxy has suffered some violent event such as an explosion in the past. One theory is that the activity in M87 is due to a massive black hole at its centre.

The Virgo Cluster is of particular interest to astronomers because it is the nearest large cluster of galaxies to us. Most galaxies, perhaps all, are grouped together in clusters of various sizes. Our own Galaxy is a member of a small cluster of 30 or so members called the Local Group, and according to some astronomers the Local Group is itself a member of a so-called supercluster centred on the Virgo Cluster. Estimates of the distance to the Virgo Cluster vary, but 40 million light years is a widely accepted figure. At this distance none of the galaxies appear particularly large or bright, but a number of them are within range of small telescopes. In addition to M87, mentioned above, look for M49, an 8th-magnitude elliptical galaxy; M60, a 9th-magnitude elliptical galaxy; M84 and M86, a pair of 9th-magnitude elliptical galaxies.

Another galaxy of note in Virgo is found on the constellation's southern border with Corvus. This is M104, an 8th-magnitude spiral galaxy that is not a member of the Virgo Cluster but which lies somewhat closer to us, 30 million light years away. Although M104 is a spiral galaxy we see it edge-on, so it actually appears elliptical when seen through small telescopes. It is popularly known as the Sombrero Galaxy, because it resembles a wide-brimmed hat on long-exposure photographs, but do not expect to see it as anything more than an elongated smudge through a small telescope.

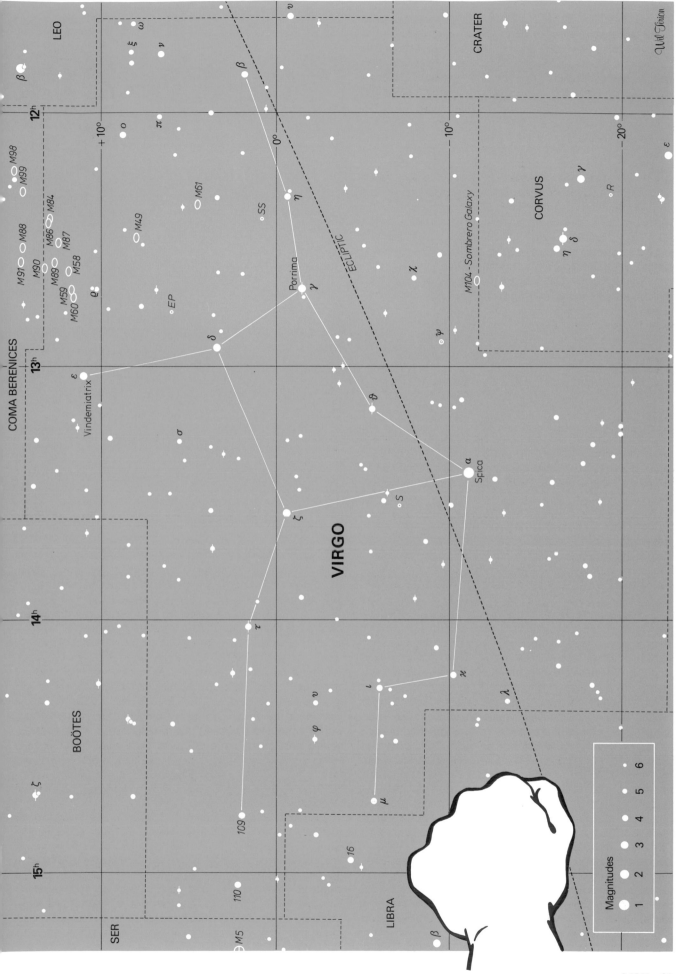

LEO

COMA BERENICES

BOÖTES

SER

VIRGO

LIBRA

CRATER

CORVUS

ECLIPTIC

12ʰ

13ʰ

14ʰ

15ʰ

+ 10°

0°

− 10°

− 20°

β

ω

ξ

ν

ν

β

π

ο

η

SS

M61

M49

δ

ε
Vindemiatrix

σ

ζ

τ

ι
ν
φ

μ

109

16

110

M5

β

M98
M99
M91
M90
M88
M86
M84
M89
M87
M58
M59
M60
ϱ
EP

Porrima
γ

ϑ

S

α
Spica

κ

λ

χ

ψ

M104 - Sombrero Galaxy

η δ

γ

R

ε

ζ

Will Tirion

Magnitudes
1 2 3 4 5 6

MAY 35

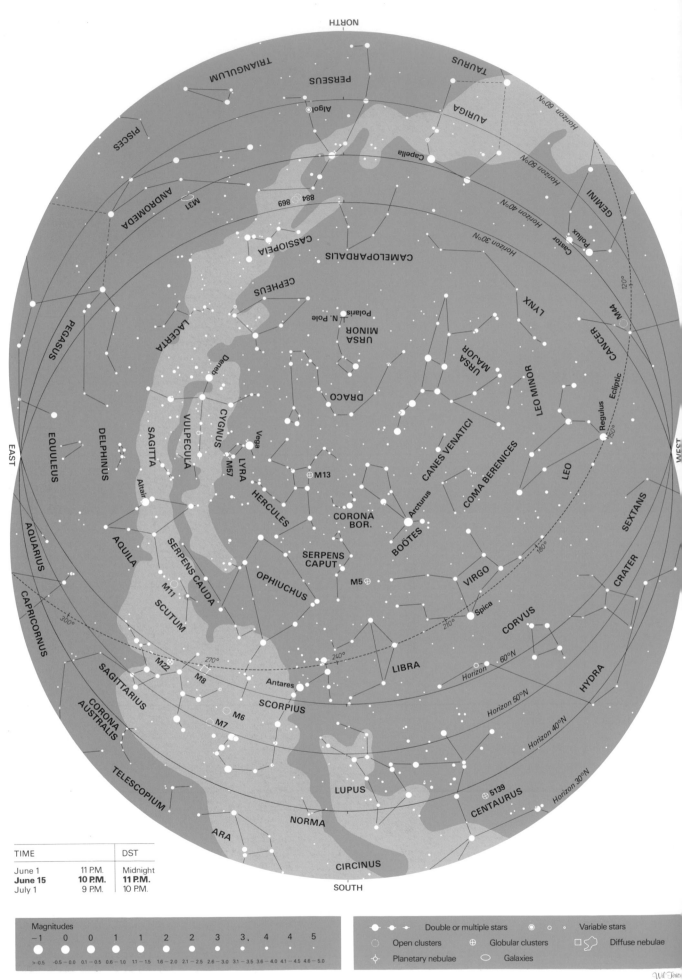

TIME		DST
June 1	11 P.M.	Midnight
June 15	**10 P.M.**	**11 P.M.**
July 1	9 P.M.	10 P.M.

Magnitudes

−1	0	0	1	1	2	2	3	3,	4	4	5
>-0.5	-0.5 - 0.0	0.1 - 0.5	0.6 - 1.0	1.1 - 1.5	1.6 - 2.0	2.1 - 2.5	2.6 - 3.0	3.1 - 3.5	3.6 - 4.0	4.1 - 4.5	4.6 - 5.0

Double or multiple stars   Variable stars
Open clusters   Globular clusters   Diffuse nebulae
Planetary nebulae   Galaxies

Wil Tirion

JUNE

The planets this month

Venus

1988 Too close to the Sun throughout most of the month, emerging into the morning sky at the end of the month.

1989 An evening object at mag. –3.3, starting the month in Taurus, crossing Gemini, and ending the month in Cancer.

1990 A morning object, moving through Aries at mag. –3.4 and ending the month in Taurus between the Hyades and Pleiades clusters.

1991 An evening object. Venus starts the month in Gemini, crosses Cancer, and ends the month in Leo. Greatest elongation (maximum separation from the Sun) is on June 13, at mag. –3.9. Overtakes the fainter Jupiter in the third week of June, and is close to Mars in the last week of June.

1992 Too close to the Sun for observation throughout the month.

Mars

1988 In Aquarius, brightening as it moves towards Pisces. Ends the month at mag. –0.6.

1989 Moves from Gemini into Cancer at mag. 2.0, ending the month just south of the Beehive Cluster (Praesepe).

1990 In Pisces at mag 0.6.

1991 A second-magnitude object, moving from Cancer into Leo, and passing across the Beehive Cluster (Praesepe) in the early part of the month. Forms a close pair with Jupiter in mid-June and with much brighter Venus around June 23.

1992 Moves from Pisces into Aries at mag. 1.1.

Jupiter

1988 Moves from Aries into Taurus, at mag. –1.6.

1989 Too close to the Sun for observation throughout the month. At conjunction (directly behind the Sun) on June 9.

1990 At mag. –1.4 in Gemini, vanishing into the evening twilight towards the end of the month.

1991 In Cancer at mag. –1.4. Overtaken by fainter Mars in mid-June and by brighter Venus the following week.

1992 In Leo at mag. –1.9.

Saturn

1988 In Sagittarius at mag. 0.2. At opposition (due south at midnight) on June 20, 1350 million km from Earth.

1989 In Sagittarius at mag. 0.3.
1990 In Sagittarius at mag. 0.4.
1991 In Capricornus at mag. 0.6.
1992 In Capricornus at mag. 0.7.

Hercules

Two giants stand head to head in the June sky. Hercules represents the Greek hero who undertook twelve labours on the orders of King Eurystheus of Mycenae. The other figure is Ophiuchus, a man entwined by a serpent. Both constellations are large – Hercules is the fifth-largest in the sky and Ophiuchus is the eleventh-largest – but neither of them is particularly prominent. Hercules contains no star brighter than third magnitude and consequently is not easy to find. Look for him between the bright stars Vega and Arcturus.

Old star maps depict Hercules as a man on one knee, brandishing a club. One foot, marked by Iota (ι) Herculis, is resting on the head of Draco, the Dragon. To the ancient Greeks this constellation was simply The Kneeling Man, 'whose name none can tell, nor what he labours at', as the Greek poet Aratos wrote in the third century BC. The identification with Hercules came later. The fact that this constellation was anonymous to the ancient Greeks suggests that they did not originate it, but inherited it from an earlier civilization, probably the Babylonians of the Middle East.

Hercules appears upside-down in the sky. His head lies in the south of the constellation, near the border with Ophiuchus, and is marked by the star Alpha (α) Herculis, popularly known as Ras Algethi, from the Arabic meaning kneeler's head. Alpha Herculis is usually classed as a red giant, but sometimes it is put among the supergiants. In common with most giant stars it changes in size, varying in brightness as it does so. It fluctuates between 3rd and 4th magnitude with no set period.

Alpha Herculis is one of the largest stars known, but exactly how big it is remains uncertain because its distance and luminosity have not been accurately determined. The star lies at least 300 light years away, and possibly twice that far. Assuming a distance of around 400 to 500 light years, it must emit more light than 1000 Suns and have a diameter about 500 times that of the Sun, so huge that it could easily swallow the orbit of the planet Mars.

Alpha Herculis is also a glorious but tight double star for small telescopes. Its close companion of 5th magnitude appears greenish by contrast with the burnished copper colour of Alpha Herculis itself. The two stars form a slow-moving binary, orbiting every few thousand years.

Hercules is a good hunting ground for double stars. Start with Delta (δ) Herculis, an unrelated duo of 3rd and 8th magnitudes. Another unrelated pair is Kappa (κ) Herculis, 'a 5th-magnitude yellow giant with a wide 6th-magnitude orange giant companion. Rho (ρ) Herculis is a neat 5th- and 6th-magnitude pair for small telescopes, both white in colour. Look next at 95 Herculis, an attractive pair of 5th-magnitude stars appearing silver and gold in colour like Christmas-tree decorations.

Finally, do not miss 100 Herculis, a matched pair of 6th-magnitude white stars, easily divisible by small telescopes. The most distinctive part of Hercules is the 'keystone' representing his pelvis, formed by four stars, Zeta (ζ), Eta (η), Epsilon (ε) and Pi (π) Herculis. Between Eta (η) and Zeta (ζ) Herculis lies one of the showpieces of the heavens, the great globular cluster M13, a city of 300,000 stars or more, compressed into a ball 100 light years across. M13 is one of 150 globular clusters that are scattered in a halo around our Galaxy. It is the brightest globular cluster in northern skies, glimpsed by the naked eye on clear nights as a luminous patch resembling a 6th-magnitude star out of focus. Binoculars show it as a mysterious glowing ball, covering about one-third the diameter of the full Moon. Note how its brightness increases towards the centre, where the stars are most crowded. The brightest stars in the cluster are giants, emitting the light of 1000 Suns. Small telescopes show those giants superimposed on the general background glow, giving the cluster an amazing granular appearance at high magnification. Anyone who lived on a planet in M13 would see the sky filled with thousands of dazzling stars, some rivalling the full Moon in brilliance. There would be no true night at all.

Hercules contains a second globular cluster, M92, smaller than M13 and about half the brightness but still easily visible through binoculars. Compare the two clusters and note that M92 has a brighter centre than M13, because its stars are more densely packed. M92 is just over 25,000 light years away, somewhat further than M13.

Ophiuchus

One of the lesser-known constellations of the sky is Ophiuchus, representing the Greek god of medicine Asclepius (the Roman Aesculapius). His symbol was the snake, and in the sky Ophiuchus is inextricably linked with the constellation of Serpens, the Serpent, which is wrapped around him. Ophiuchus divides Serpens into two, an arrangement which is unique among the constellations. Serpens Caput, the serpent's head, is on one side of Ophiuchus, while Serpens Cauda, the tail, lies on the other side. Nevertheless, the two halves of Serpens are regarded as one constellation. This stellar triptych of Ophiuchus and the two halves of Serpens lies in the south during summer nights.

Ophiuchus contains a number of globular clusters, the brightest being M10, M12 and M62. All three, of 7th magnitude, are visible in binoculars, but they pale by comparison with M13 in the Hercules. Far more appealing is the open cluster IC 4665, a scattering of 20 or so stars of 7th magnitude and fainter, in the same field as the orange giant Beta (β) Ophiuchi. This cluster is ideal for binoculars because of its large size, greater than that of the full Moon.

For an attractive double star, turn a small telescope on 36 Ophiuchi in the constellation's southern reaches and see a neat pair of matching 5th-magnitude orange dwarf stars. More challenging is 70 Ophiuchi, a beautiful golden-yellow and orange duo, magnitudes 4.2 and 6.0, that orbit each other every 88 years. As seen from Earth, the two are closest together around the year 1990 at a separation of 1.5 seconds of arc, needing a telescope of at least 75 mm aperture to resolve. Thereafter the two stars move apart again, becoming divisible with a 60 mm aperture by 1993, and with 50 mm in 1995, although a steady night and high magnification will be needed. It should be interesting to watch these two stars become progressively easier to split in small telescopes year by year as their separation increases. No observer should miss 70 Ophiuchi, one of the most famous double stars in the sky.

On the constellation's border with Scorpius, just north of Antares, lies an outstanding multiple star for small apertures, Rho (ρ) Ophiuchi. Low magnification shows a V-shaped grouping consisting of a 5th-magnitude star at the apex, accompanied by two stars of 7th magnitude. High powers show that the star at the triangle's apex is itself double, consisting of a tight pair of 5th and 6th magnitudes. Long-exposure photographs show that this whole area is enveloped in a faint haze of nebulosity that extends southwards to Antares.

BARNARD'S STAR

Oddly enough, the most celebrated star in Ophiuchus is not visible to the naked eye. Nor is it easy to find in small telescopes. Yet it is worth hunting for, because it is the second-closest star to the Sun, the fastest-moving star, and it may be the centre of the nearest planetary system to our own.

It is Barnard's Star, a red dwarf of magnitude 9.5, six light years away, named after the American astronomer Edward Emerson Barnard. In 1916, Barnard noticed that this faint star had changed markedly in position on two photographs taken 22 years apart. All stars are on the move, but most of them move almost imperceptibly. Barnard's Star has the fastest movement across the sky of any star, changing in position by 10.3 seconds of arc per year. In 180 years it crosses the apparent diameter of the full Moon.

In addition to its movement across the sky, Barnard's Star is also approaching the Sun. It will reach its minimum distance of 3.8 light years from us in AD 11,800, closer even than Alpha Centauri is today. Yet it is so feeble that even then it will appear of only magnitude 8.5. Barnard's Star has a mass about 15 per cent that of the Sun, and is 2300 times fainter than the Sun.

According to studies by Peter van de Kamp of Sproul Observatory, Pennsylvania, Barnard's Star may be accompanied by two planets with masses similar to those of Jupiter and Saturn. These planets cannot be seen directly through telescopes, but their presence is inferred by a wobble they cause in the star's motion across the sky.

Barnard's Star lies to the left of Beta (β) Ophiuchi, near the 5th-magnitude star 66 Ophiuchi, which is slightly variable. The chart shows how rapidly the star's position moves. Observe it over a period of years so that you, too, can spot the motion of this runaway star.

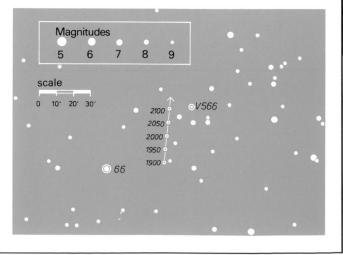

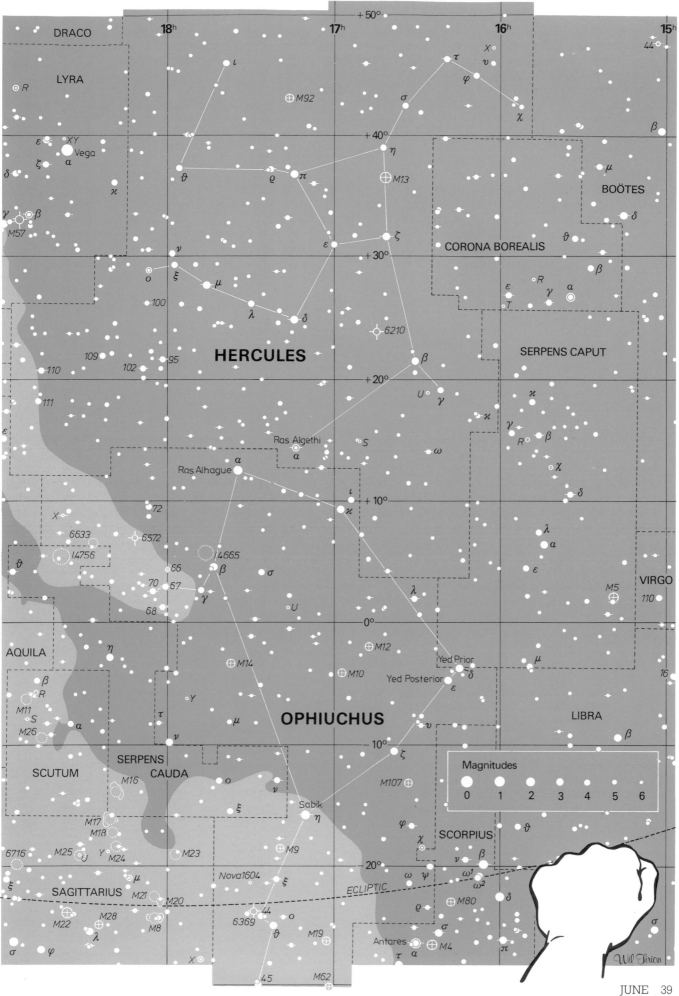

DRACO

LYRA

⊙ R

ε XY
Vega
ζ
α ϰ

δ
γ ⊙ β
M57
ν

ξ
ο
100

109
110
102
95

111

ε

18ʰ
ι

⊕ M92

ϑ
ϱ
π

μ
λ
δ

HERCULES

17ʰ
+50°

τ χ
ν
φ

σ

+40°
η

⊕ M13
ζ

ε

+30°

◇ 6210

β
+20°
U ⊙
γ

16°
44

β

μ
BOÖTES
δ

ϑ

β

CORONA BOREALIS

ε R
γ α ⊙
T

SERPENS CAPUT

ϰ

ϰ
γ
R ⊙ β

⊙ χ

δ

15ʰ

Ras Algethi
α

S

ω

VIRGO

Ras Alhague α
72

6633
6572
14756

X
ϑ

66
70 57
68

14665
β

γ

σ

U

ι
ϰ
+10°

λ
α
ε

λ
M5
⊙ 110

AQUILA
β
R
M11
S
M26
α

η

τ
ν

μ

ν

0°

⊕ M14

Y

OPHIUCHUS

μ

⊕ M12

⊕ M10

Yed Prior
δ
Yed Posterior
ε

ν

μ

LIBRA
β

16°

SCUTUM

SERPENS
CAUDA

M16
ο
ν
ξ

Sabik
η

-10°
ζ

M107 ⊕

φ
χ

SCORPIUS
ϑ

Magnitudes
0 1 2 3 4 5 6

6716
M25 U
Y M24

M17
M18

M23

⊕ M9

Nova1604
ξ

SAGITTARIUS
μ
M21
M20

M22 M28
λ M8

6369
44
ο
ϑ

M19
⊕

ECLIPTIC

ω ψ
ω¹
ω²
δ

ϱ ⊕ M80

σ

σ
φ

X ⊙

45 M62

-20°

Antares
τ α ⊙ ⊕ M4
π

Wil Tirion

NORTH

AURIGA
GEMINI

Horizon 60°N

Horizon 50°N

PERSEUS
Capella

Algol
Horizon 40°N
CAMELOPARDALIS
LYNX

ARIES
TRIANGULUM
ANDROMEDA
Horizon 30°N
884
869

CANCER

PISCES
M31
CASSIOPEIA
CEPHEUS
URSA MINOR
N. Pole
Polaris
URSA MAJOR
CANES VENATICI
LEO MINOR

LEO

EAST
PISCES
LACERTA
DRACO
COMA BERENICES
WEST

PEGASUS
CYGNUS
Deneb
LYRA
Vega
M13
Arcturus
180°

DELPHINUS
SAGITTA
VULPECULA
M57
CORONA
BOR.
BOÖTES

EQUULEUS
HERCULES
SERPENS CAPUT
VIRGO

320°
AQUILA
Altair
SERPENS
CAUDA
M5
Spica

AQUARIUS
Ecliptic
OPHIUCHUS
270°

PISCIS
AUSTRINUS
M11
SCUTUM
CORVUS

300°
HYDRA

CAPRICORNUS
M22
240°
LIBRA
Horizon 60°N

270°
M8
Antares
Horizon 50°N

SAGITTARIUS
M6
SCORPIUS
CENTAURUS

GRUS
MICROSCOPIUM
M7
Horizon 40°N
Horizon 30°N

CORONA
AUSTRALIS
LUPUS

INDUS
TELESCOPIUM
NORMA

ARA

SOUTH

TIME		DST
July 1	11 P.M.	Midnight
July 15	**10 P.M.**	**11 P.M.**
August 1	9 P.M.	10 P.M.

Magnitudes
−1   0   0   1   1   2   2   3   3   4   4   5
>-0.5 -0.5–0.0 0.1–0.5 0.6–1.0 1.1–1.5 1.6–2.0 2.1–2.5 2.6–3.0 3.1–3.5 3.6–4.0 4.1–4.5 4.6–5.0

Double or multiple stars   Variable stars
Open clusters   Globular clusters   Diffuse nebulae
Planetary nebulae   Galaxies

Wil Tirion

JULY

The planets this month

Venus
1988 Brilliant in the morning sky at mag. −4.2, in Taurus.
1989 Moves from Cancer into Leo in the evening sky at mag. −3.3, overtaking Mars on July 12 and passing close to Regulus on July 23.
1990 In the morning sky, moving from Taurus into Gemini at mag. −3.3.
1991 A brilliant evening object in Leo at mag. −4.2, far outshining Regulus, which it passes on July 12. Venus and Mars pass each other in the last week of July.
1992 Lost in evening twilight throughout the month.

Mars
1988 In Pisces, brightening markedly during the month from mag. −0.7 to mag. −1.3 as it approaches the Earth.
1989 Moves from Cancer into Leo at mag. 2.0, ending the month near Regulus (mag. 1.4). Bright Venus (mag. −3.3) passes on July 12.
1990 Moves from Pisces into Aries, brightening from mag. 0.5 to mag. 0.2.
1991 In Leo at mag. 2.0, passing Regulus on July 14 and Venus on July 22.
1992 Starts the month as a first-magnitude object in Aries then moves into Taurus, ending the month between the Hyades and Pleiades clusters.

Jupiter
1988 In Taurus, south of the Pleiades cluster.
1989 Starts the month in the morning twilight in Taurus at mag. −1.5, ends the month by crossing the border into Gemini.
1990 Too close to the Sun for observation throughout the month. At conjunction (directly behind the Sun) on July 15.
1991 Moves from Cancer into Leo at mag. −1.3, but moves into the evening twilight by the end of the month.
1992 In Leo at mag. −1.3.

Saturn
1988 In Sagittarius, near the border with Ophiuchus at mag. 0.4.
1989 In Sagittarius at mag. 0.2. At opposition (due south at midnight) on July 2, 1350 million km from Earth.
1990 In Sagittarius at mag. 0.3. At opposition (due south at midnight) on July 14.
1991 In Capricornus at mag. 0.3. At opposition (due south at midnight) on July 27.
1992 In Capricornus at mag. 0.5.

Lyra

Overhead this month lies the compact but distinctive constellation of Lyra, the lyre, which represents the harp of the Greek bard Orpheus. In legend, Hermes made the harp from a tortoise shell and gave it to his half-brother Apollo. In turn, Apollo passed the harp to Orpheus who, it was said, could charm wild beasts and even trees with his musical skill. This ability was particularly valuable when Orpheus entered the fearful Underworld to plead for the return of his dead wife, Eurydice. Hades, god of the Underworld, was so moved by the music of Orpheus that he agreed to release Eurydice to the land of the living, on condition that Orpheus did not look back at her as he led her to the upper world. But, just before they emerged into daylight, Orpheus glanced round to be sure she was following, and lost her forever. Thereafter he roamed the world in sorrow, playing his harp plaintively. After Orpheus died, his harp was placed in the heavens in commemoration of his musical gifts.

The constellation has also been identified as an eagle or vulture, often depicted on early star maps carrying the harp. Whatever your interpretation of Lyra, it has a rich stock of celestial attractions to charm astronomers.

Lyra is easily located by its brightest star Alpha (α) Lyrae, better known as Vega, the fifth-brightest star in the entire sky. Vega's name comes from the Arabic meaning swooping eagle, for the Arabs were among those who visualized the constellation as an eagle. Vega is a blue-white star of magnitude 0.03, 26 light years away. It has a mass of about three Suns and gives out 50 times as much light as the Sun. In 1983 the Infra-Red Astronomy Satellite, IRAS, made the astounding discovery that Vega is surrounded by a disk of cold, dark dust that is forming into a system of planets – another solar system. Vega is a much younger star than the Sun, with an age of only a few hundred million years. Perhaps, in billions of years, life will evolve on one of those planets that are now forming around Vega.

Near to Vega, binoculars or even sharp eyesight shows a wide 5th-magnitude double star, Epsilon (ε) Lyrae, which has a delightful surprise awaiting the users of telescopes. Both stars are themselves double, forming a spectacular stellar quadruplet that is popularly known as the Double Double.

Epsilon-1, the slightly wider pair, consists of stars of magnitudes 5.0 and 6.1 orbiting every 1200 years or so. Currently they are closing together and will continue to do so through the twenty-first century. The predicted separations are 2.6 seconds of arc in the year 2000 and 2.5 seconds in 2020. Stars with such separation should in theory be divisible by an aperture of 50 mm, but the brightness difference means that in practice at least 60 mm will be required.

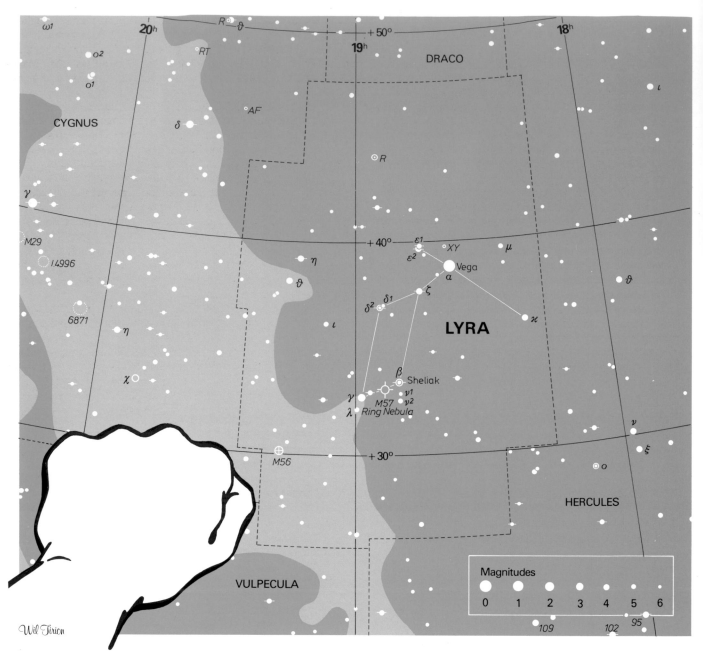

Epsilon-2 Lyrae, the closer pair, consists of stars of magnitudes 5.2 and 5.5 whose orbit lasts over 500 years. Their separation is increasing slightly, reaching 2.3 arc seconds in 2000 and 2.4 in 2020, but they will still be difficult to split in 50 mm telescopes. Whatever your telescope, do not miss this showpiece object, the finest quadruple star in the sky.

Epsilon Lyrae is not the only double star in this constellation with a surprise up its sleeve. Look at Beta (β) Lyrae, an attractive coupling of a cream primary with an 8th-magnitude blue star, easily separated by small telescopes. But the primary star is itself a complex eclipsing binary that varies between magnitudes 3.3 and 4.3 every thirteen days. Compare its brightness with that of Gamma (γ) Lyrae, constant at magnitude 3.2. Astronomers have struggled for decades to understand the curious primary of Beta Lyrae, whose period of variation is gradually lengthening. The answer has come from carefully studying the star's light through spectroscopes. Evidently the two eclipsing stars are so close together (closer than Mercury is to the Sun) that each is distorted into an egg shape by the other's gravity. Gas flows between the stars, some of it spiralling away into space. When you look at Beta Lyrae, try to imagine the turbulent activity taking place there, unseen by the human eye.

Another double-variable is Delta (δ) Lyrae, an easy binocular duo consisting of a 4th-magnitude red giant star that is slightly variable and a blue-white star of magnitude 5.6. The two lie at different distances from us and so are unrelated. Look also at Zeta (ζ) Lyrae, a binocular duo of magnitudes 4.4 and 5.7.

Between Beta and Gamma Lyrae lies another of the treasures of Lyra, the famous Ring Nebula, M57, a celestial smoke ring thrown off by a dying star about 2000 light years away. M57 is a type of object known as a planetary nebula, not because it has anything to do with planets but because it shows a rounded disk like a planet when viewed through a telescope. Such an object is believed to be formed when a star like the Sun reaches the end of its life. After swelling up into a red giant, the star sloughs off its distended outer layers to form a planetary nebula, leaving the core as a faint white dwarf.

Small telescopes show M57 as a 9th-magnitude hazy spot, larger than the disk of Jupiter. Its ring shape becomes apparent in apertures of at least 100 mm, which also reveal its elliptical outline. Its central star, of 15th magnitude, is too faint for amateur telescopes. Unfortunately, M57 is not as impressive visually as it is on long-exposure photographs.

Vulpecula

Adjoining Lyra is Vulpecula, the fox, a little-known constellation well worth an introduction. Like its southerly neighbour Sagitta, Vulpecula straddles a rich area of the Milky Way and is a good hunting ground for novae. The constellation was placed in the sky in 1690 by the Polish astronomer Johannes Hevelius under the title of Vulpecula cum Anser, the fox and goose, which has since been shortened. Hevelius said that he placed a fox here to be close to two other predatory animals, the vulture (Lyra) and the eagle (Aquila).

Vulpecula achieved unexpected fame in 1967 when radio astronomers at Cambridge, UK, discovered the first of the flashing radio stars called pulsars south of Alpha (α) Vulpeculae. Pulsars are thought to be the highly condensed cores of stars that have exploded as supernovae. Like most pulsars, this one is so faint that it is invisible optically.

Close to the border with Sagitta lies one of the oddest-looking star groupings in the sky, sometimes known either as Brocchi's Cluster or Collinder 399 but more commonly termed the Coathanger because of its distinctive shape. Its main feature is an almost perfectly straight line of six stars, from the centre of which protrudes a curving hook of four more stars, completing the coathanger shape. The ten stars of the Coathanger are from 5th to 7th magnitude, and the whole object is 1.5 degrees long, making it an ideal binocular sight.

The Coathanger may not be a true cluster, because data on its individual stars suggest that they lie at a wide range of distances, from about 300 light years to over 4000 light years. If these distances are correct, the stars of the Coathanger cannot be related, and lie in the same line of sight only by chance. Whatever the case, the Coathanger remains a delightful surprise.

From the Coathanger, turn your binoculars to the best-known object in Vulpecula, M27, popularly known as the Dumb-bell Nebula. M27 is a planetary nebula, a shell of gas thrown off by a dying star, like the Ring Nebula in Lyra. But the Ring Nebula and the Dumb-bell have noticeably different appearances.

Whereas many planetary nebulae are difficult to spot because they are so small, the Dumb-bell can be missed because it is unexpectedly large. At 8th magnitude the Dumb-bell is bright enough to show up in binoculars as a smudgy patch even under indifferent sky conditions. It is elongated, with a maximum diameter about a quarter that of the full Moon, eight times larger than the Ring Nebula. Having a favourable combination of size and brightness, M27 is the most conspicuous of all planetary nebulae, at least in small instruments.

Seen through a telescope, the Dumb-bell is greenish in colour. The Dumb-bell gets its name from its double-lobed shape in telescopes, reminiscent of an hour-glass, or perhaps a figure eight, clearly different from the shape of the Ring Nebula. Its central star, the white dwarf which has shed its outer layers to form the nebula, is of 14th magnitude, too faint for small telescopes. The Dumb-bell lies about 1000 light years away, among the closest planetary nebulae to us.

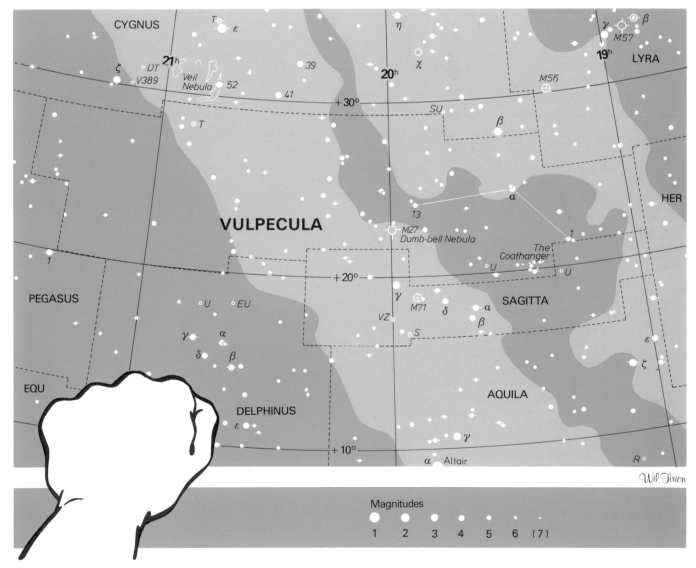

Wil Tirion

Magnitudes

1 2 3 4 5 6 (7)

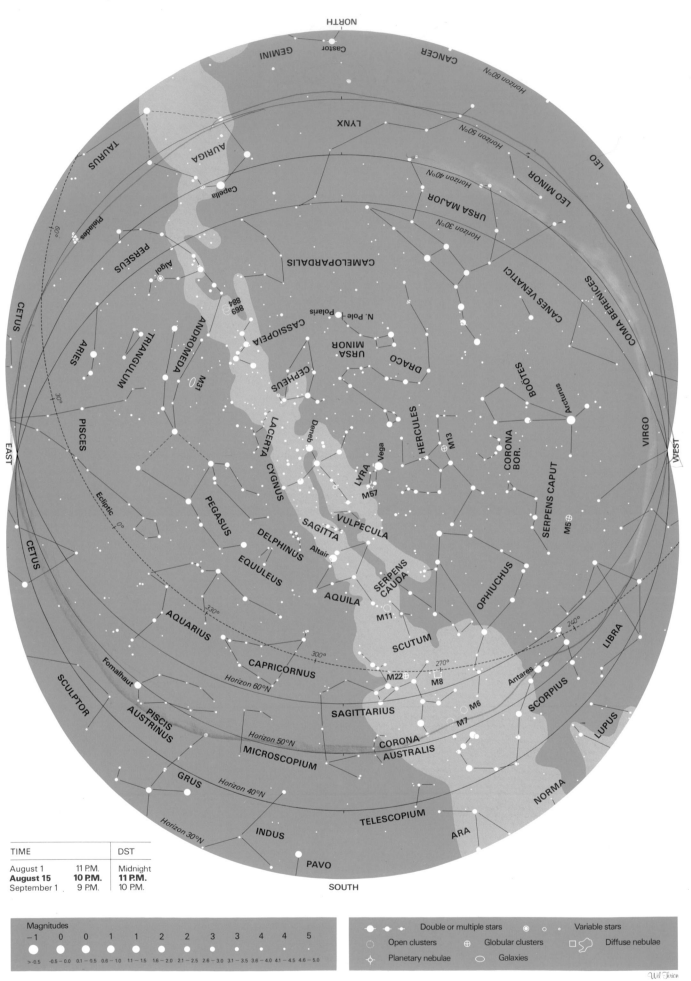

TIME		DST
August 1	11 P.M.	Midnight
August 15	**10 P.M.**	**11 P.M.**
September 1	9 P.M.	10 P.M.

Magnitudes

−1	0	0	1	1	2	2	3	3	4	4	5
> -0.5	-0.5 – 0.0	0.1 – 0.5	0.6 – 1.0	1.1 – 1.5	1.6 – 2.0	2.1 – 2.5	2.6 – 3.0	3.1 – 3.5	3.6 – 4.0	4.1 – 4.5	4.6 – 5.0

Double or multiple stars   Variable stars

Open clusters   Globular clusters   Diffuse nebulae

Planetary nebulae   Galaxies

Wil Tirion

AUGUST

The planets this month

Venus

1988 A brilliant morning object, moving from Taurus into Gemini. Greatest elongation (maximum separation from the Sun) is on August 22, when Venus is at mag. −4.0.

1989 An evening object at −3.4, moving from Leo into Virgo.

1990 A morning object, moving from Gemini through Cancer and into Leo at mag. −3.3, ending the month in the dawn twilight. Passes Jupiter (mag. −1.4) on August 13.

1991 In the evening sky, south of the ecliptic in Sextans. Starts the month at mag. −4.0 but fades and vanishes into the evening twilight by mid-month.

1992 Emerges from twilight into the evening sky early in the month. Starts the month in Leo, passing Regulus on August 6 and Jupiter on August 23. Ends the month in Virgo.

Mars

1988 South of the ecliptic in Cetus, brightening impressively from mag. −1.3 to −2.0 during the month.

1989 Starts the month in Leo just north of Regulus, and fainter at mag. 2.0. Engulfed by evening twilight after the first week of August.

1990 Moves from Aries into Taurus, brightening from mag. 0.2 to −0.2, finishing the month south of the Pleiades.

1991 Moves from Leo into Virgo at mag 2.0.

1992 In Taurus. Starts the month between the Hyades and Pleiades clusters, then moves towards the border with Gemini, brightening to mag. 0.8.

Jupiter

1988 In Taurus, between the Pleiades and Hyades clusters, at mag. −1.9.

1989 At the feet of Gemini, mag. −1.6.

1990 Emerges from the morning twilight after the first week of August at mag. −1.4. Passes Venus (mag. −3.3) on August 13.

1991 Too close to the Sun to be seen throughout the month. At conjunction (directly behind the Sun) on August 17.

1992 In Leo at mag. −1.2. Passed by Venus on August 23. Disappears into the evening twilight at the end of the month.

Saturn

1988 In Sagittarius, on the border with Ophiuchus at mag. 0.5.

1989 In Sagittarius at mag. 0.4.

1990 In Sagittarius at mag. 0.4.

1991 In Capricornus at mag. 0.4.

1992 In Capricornus at mag. 0.4. At opposition (due south at midnight) on August 7, 1330 million km from Earth.

August meteors

Warm summer nights are ideal for meteor observing, and August has the brightest of the year's meteor showers, the Perseids. This is the shower on which to begin meteor observing. The Perseids reach a peak around August 12 every year, when as many as one meteor a minute may be seen streaking away from the northern part of the constellation Perseus. The radiant lies close to Gamma (γ) Persei, but it does not rise very high until late at night, so do not expect to see many Perseids before midnight. Continue watching into the early morning hours, for the rate of Perseids should increase towards dawn as the radiant climbs higher in the sky. Perseids meteors are spectacular. They are usually bright, they frequently flare up and many of them leave glowing trains. The shower is very broad, so that considerable numbers of Perseids can be seen for over a week before and after maximum. They move along the same orbit as comet Swift-Tuttle, which was discovered in 1862 but has not been seen since.

Cygnus

Everyone has heard of the Southern Cross, as the southern constellation Crux is popularly called, but it is less well known that there is a far larger cross in the northern sky. The proper name for this constellation is Cygnus, the Swan, but it is often termed the Northern Cross.

The Greeks said that Cygnus represented the swan into which the god Zeus transformed himself in order to seduce Leda, wife of King Tyndareus of Sparta. On summer nights, Cygnus can be seen flying along the sparkling band of the Milky Way, its long neck outstretched towards the south-west and its stubby wings tipped by the stars Delta (δ) and Epsilon (ε) Cygni.

The swan's tail is marked by Alpha (α) Cygni, better known as Deneb, a name that comes from the Arabic for tail. Deneb, of magnitude 1.3, is the faintest of the three stars that make up the Summer Triangle. But whereas the other two stars of the Triangle, Vega and Altair, are among the closest stars to the Sun, Deneb lies far off, nearly 2000 light years away. It is the most distant first-magnitude star, twice as far as Rigel which holds second place.

To be easily visible over such an immense distance, Deneb must be exceptionally powerful. It is in fact among the most luminous of stars, a supergiant with a white-hot surface giving out over 80,000 times as much light as the Sun. If Deneb were as close to us as the nearest star, Sirius, it would appear as bright as a half Moon, so the sky would never be dark when it was above the horizon.

Even more remarkable is the star P Cygni, located near the base of the swan's neck. It is currently of 5th magnitude, but on two occasions during the seventeenth century it temporarily brightened to 3rd magnitude. P Cygni is evidently so massive that it is unstable, throwing off shells of gas at irregular intervals, causing its surges in brightness. It will probably end its life by exploding as a supernova. Although data about such peculiar stars are highly uncertain, P Cygni is thought to have a mass of about 50 Suns and to shine as brightly as 100,000 Suns. It lies about three times farther away than Deneb, among the distant stars of the Milky Way.

Cygnus is an ideal region to explore on warm summer evenings. Begin by sweeping with binoculars along the star-splashed Milky Way, particularly rich in this area. The star clouds in Cygnus are part of one of the spiral arms of our Galaxy, over 6000 light years distant. Look for a major division in the Milky Way called the Cygnus Rift, caused by a dark lane of dust which blocks starlight from behind.

Of the star clusters in Cygnus, the best in binoculars is M39, a scattered group of 20 or so stars of 7th magnitude and fainter, arranged in a triangular shape. M39 lies about 900 light years away, much closer than the background stars of the Milky Way.

Under clear, dark skies binoculars should show the large, misty shape of the North America Nebula, so-called because of its remarkable resemblance to the continent of North America, although its shape is more apparent on long-exposure photographs than it is through small instruments. The Nebula, also known as NGC 7000, lies near the star Xi (ξ) Cygni, and covers three times the width of the Moon, so large that only binoculars or wide-angle telescopes can encompass it all.

This immense mass of gas and dust, nearly 50 light years in diameter, lies at a similar distance to the star Deneb. It has often been suggested that the North America Nebula is lit up by Deneb, but the separation between star and Nebula is considerable, and other stars that lie within the Nebula may instead be responsible for making it glow.

While scanning Cygnus, be sure to look at Omicron-1 (o^1) Cygni, a binocular gem. This orange star, magnitude 3.8, forms a beautiful wide pairing with the turquoise 30 Cygni, exactly one magnitude fainter. Binoculars, if held steadily, will also reveal a 7th-magnitude blue star closer to Omicron-1 Cygni.

This is an ideal warm-up before turning your gaze upon the queen of double stars, the beautiful Beta (β) Cygni or Albireo. (Its name, incidentally, results from a mis-translation and is completely meaningless.) Albireo is easily found, marking the head of the swan – or, if you prefer, the foot of the cross. The smallest of telescopes, and even binoculars mounted steadily, show that Albireo consists of a glorious pair of amber and blue-green stars, magnitudes 3.1 and 5.1, like a celestial traffic light. These beautifully contrasting colours, similar to those of Omicron-1 Cygni, will appear more intense if you put the stars slightly out of focus, or if you tap the telescope so that the image vibrates.

Most measurements place the two components of Albireo at the same distance, 400 light years from us. That being so, the stars are probably related, orbiting each other every 100,000 years or so. But it has also been suggested that the fainter star, the blue-green one, is up to 50 per cent farther away than the orange star, in which case the pair would be merely an optical double, i.e. they would lie in the same line of sight purely by chance. Whatever the case, Albireo remains one of the greatest attractions on any Grand Tour of the sky.

For another celebrated double, look above the crossbar of Cygnus to find 61 Cygni, a pair of orange stars of 5th and 6th magnitude, easily separated in small telescopes. The two stars orbit each other every 650 years. Considerable historical interest surrounds 61 Cygni, for it was the first star to have its distance measured by the technique of parallax. In this technique, a star's position is accurately noted on two occasions six months apart, when the Earth is on opposite sides of its orbit around the Sun. The slight shift in the star's position as seen from the two vantage points reveals how far away the star is. A nearby star will seem to move a lot, whereas a distant star will hardly move at all.

The observations of 61 Cygni were made by the German astronomer Friedrich Wilhelm Bessel in 1838. He concluded that 61 Cygni is 10.3 light years away. That is not too different from the modern value of 11.1 light years. Of the naked-eye stars, only Alpha Centauri, Sirius, and Epsilon Eridani are closer than 61 Cygni to us.

A less well-known treasure of Cygnus is a remarkable planetary nebula, NGC 6826, popularly known as the Blinking Planetary because it appears to blink on and off as you look at it and away from it. To find it, first locate 16 Cygni, an attractive duo of 6th-magnitude yellow stars easily split by small telescopes. The Blinking Planetary lies less than a degree away from this pair, so that both objects will just fit into the same telescopic field of view.

In small telescopes with low magnification NGC 6826 looks like a fuzzy, faint star. Higher magnification reveals a light-blue disk with a 10th-magnitude star at its centre, although this central star will probably be beyond the reach of the smallest telescopes. Like many delicate objects, the nebula is best seen with averted vision, that is, by looking to one side of it. Alternately looking at the nebula and away again produces the uncanny blinking effect which gives this object its popular name. Needless to say, the blinking is purely an optical effect in the observer's eye, and does not originate in the nebula itself.

Some of the most amazing objects in Cygnus are, unfortunately, beyond the reach of amateur instruments. On the border of Cygnus with Vulpecula lies an enormous loop of gas, the Veil Nebula, the twisted wreckage of a star that exploded as a supernova some 50,000 years ago. The Veil Nebula covers 3 degrees of sky, equivalent to the width of two fingers held at arm's length. The brightest portions of the Veil may just be made out in binoculars under exceptional conditions, but most observers will have to content themselves with studying photographs of it in books.

Between Gamma (γ) and Delta (δ) Cygni, in one of the wings of the swan, lies Cygnus A, one of the most intense sources of radio noise in the sky. Photographs show that it is a peculiar-looking galaxy of 15th magnitude, either exploding or undergoing a collision. It lies deep in the Universe, about 750 million light years outside the Milky Way.

Of all the objects in Cygnus the most bizarre is Cygnus X-1, the best candidate in the Milky Way for a black hole. Although the black hole itself is invisible, it gives itself away by sucking in hot gas from a neighbouring 9th-magnitude star. Gas falling towards the black hole heats up, emitting X-rays which are detected by satellites. Cygnus X-1 lies about 8000 light years away, in the Cygnus spiral arm of our Galaxy. Its position is about half a degree (one Moon's breadth) from Eta (η) Cygni.

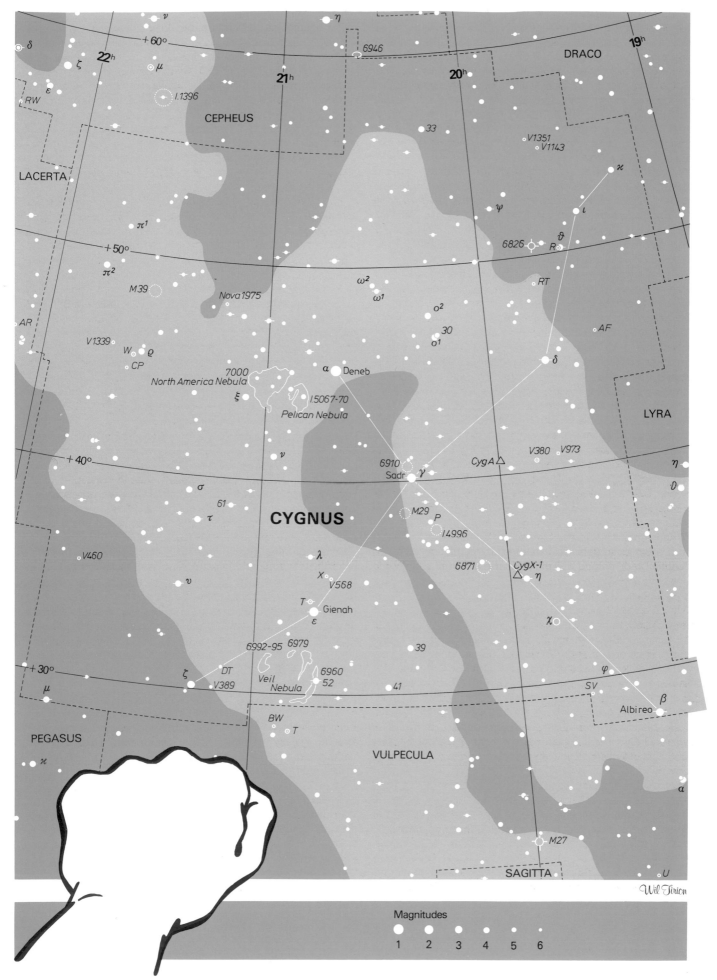

δ
ζ
ε
RW

22ʰ +60°
μ
ν
η
6946
19ʰ
21ʰ
20ʰ
DRACO
I.1396
33
V1351
V1143
CEPHEUS
ж
LACERTA
ι
ψ
ϑ
6826
R
π¹
+50°
RT
π²
AF
M39
Nova 1975
ω²
ω¹
o²
30
o¹
AR
δ
V1339
W
ϱ
CP
7000
α Deneb
North America Nebula
ξ
LYRA
I.5067-70
Pelican Nebula
η
ϑ
ν
V380 V973
+40°
6910
γ
CygA △
Sadr
M29
P
σ
14996
61
τ
CYGNUS
6871
CygX-1 △ η
V460
λ
V568 X
χ
ν
T
ε Gienah
6992-95 6979
39
6960
ζ DT
52
41
φ
V389 Veil
SV
Nebula
μ
β
BW
Albireo
PEGASUS
T
ж
VULPECULA
α
M27
U
SAGITTA

Wil Tirion

Magnitudes
1   2   3   4   5   6

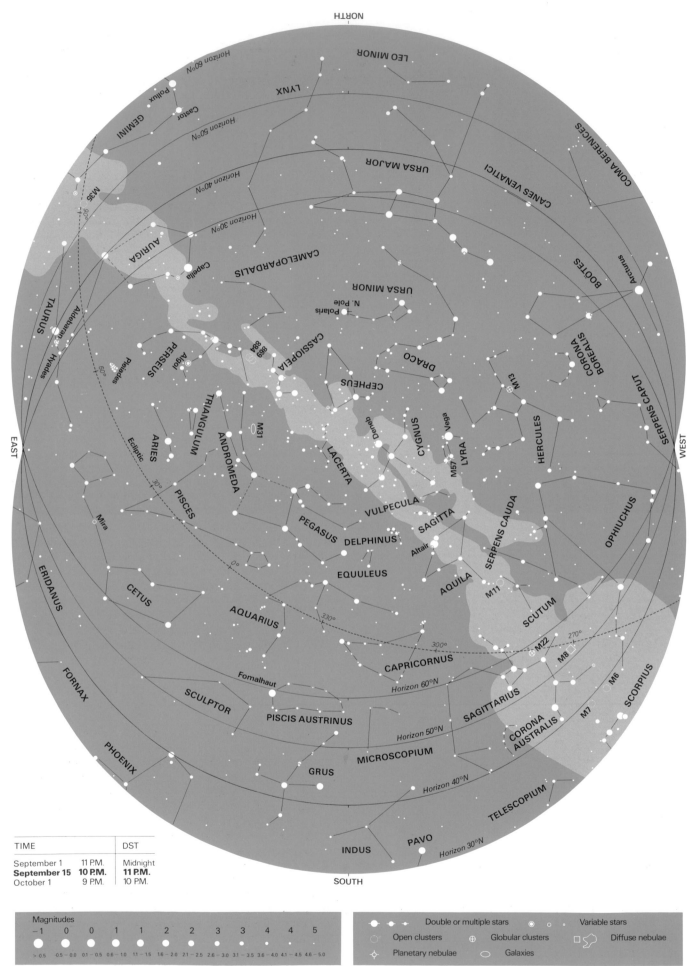

TIME		DST
September 1	11 P.M.	Midnight
September 15	**10 P.M.**	**11 P.M.**
October 1	9 P.M.	10 P.M.

Magnitudes

−1	0	0	1	1	2	2	3	3	4	4	5
> 0.5	−0.5–0.0	0.1–0.5	0.6–1.0	1.1–1.5	1.6–2.0	2.1–2.5	2.6–3.0	3.1–3.5	3.6–4.0	4.1–4.5	4.6–5.0

Double or multiple stars   Variable stars

Open clusters   Globular clusters   Diffuse nebulae

Planetary nebulae   Galaxies

SEPTEMBER

The planets this month

Venus

1988 A morning object at mag. –3.8. Starts the month in Gemini, then crosses through Cancer and ends the month in Leo.

1989 An evening object at mag. –3.5, moving from Virgo in Libra. Passes Spica on September 6.

1990 Becomes lost in the morning twilight at the start of the month.

1991 Emerges from twilight into the morning sky in Leo at the start of the month. Brightens to mag. –4.3 by the month's end.

1992 In the evening sky in Virgo at mag. –3.3, ending the month on the border with Libra. Passes Spica on September 19.

Mars

1988 On the Pisces–Cetus border. Starts the month at mag. –2.1 and brightens as it spproaches opposition (due south at midnight) on September 28, 59 million km from Earth, when it is mag. –2.5.

1989 Too close to the Sun throughout the month. Conjunction (directly behind the Sun) on September 29.

1990 In Taurus, between the Hyades and Pleiades clusters. Brightens during the month from mag. –0.3 to –0.8.

1991 A second-magnitude object in Virgo. Vanishes into the evening twilight in the second half of the month.

1992 Moves from Taurus into Gemini, brightening from mag. 0.7 to 0.5.

Jupiter

1988 In Taurus, between the Hyades and Pleiades clusters, at mag. –2.1

1989 In Gemini at Mag. –1.7.

1990 In Cancer at Mag. –1.5, passing south of Praesepe (the Beehive Cluster).

1991 Emerges from twilight into the morning sky during the second week of September. In Leo at mag. –1.3, near Regulus.

1992 Too close to the Sun to be seen throughout the month. Conjunction (directly behind the Sun) on September 17.

Saturn

1988 In Sagittarius, on the border with Ophiuchus at mag. 0.7.

1989 In Sagittarius at mag. 0.6.

1990 In Sagittarius at mag. 0.6.

1991 In Capricornus at mag. 0.5.

1992 In Capricornus at mag. 0.6.

Cassiopeia

The only married couple among the constellations are Cepheus and Cassiopeia. In Greek mythology they were King and Queen of Ethiopia, parents of Andromeda. According to fable, Cassiopeia once boasted that she was more beautiful than the group of sea nymphs called the Nereids. This incurred the wrath of the sea god Poseidon, who sent a terrible monster to ravage the lands of King Cepheus. To save his country, Cepheus was forced to offer his daughter Andromeda as a sacrifice to the monster, although she was rescued from the monster's jaws by Perseus.

Cassiopeia and Cepheus are to be found next to each other in the north polar region of the sky. Cassiopeia, the more prominent of the husband-and-wife pair, is depicted sitting in a chair. She circles incongruously around the pole, appearing to hang upside down for part of the night, a lesson in humility decreed by the Nereids.

Cassiopeia is one of the easiest constellations to recognize because of its distinctive W-shape, representing the chair in which the Queen sits. The constellation is easily located, opposite the celestial pole from the Plough. Gamma (γ) Cassiopeiae, the central star of the W-shape, is worth keeping an eye on for its unpredictable brightness, which has fluctuated between magnitudes 1.6 and 3.0 since early this century, when its variability was first noted. Compare it with neighbouring Alpha (α) Cas, magnitude 2.2, and Delta (δ) Cas, magnitude 2.7. Gamma Cassiopeiae is a hot, blue star rotating so rapidly that it throws off shells of gas, which produces the changes in brightness. As with all such peculiar stars data are uncertain, but Gamma Cas is thought to give out as much light at 5000 Suns and to lie almost 800 light years away.

A beautiful double star for small telescopes is Eta (η) Cassiopeiae, consisting of yellow and reddish stars, magnitudes 3.4 and 7.5. These two stars orbit each other every 480 years. As seen from Earth they were closest together in 1889 and will continue to move apart until the middle of the twenty-second century, making the double progressively easier to divide. The pair are only 19 light years from the Sun, relatively close on the interstellar scale.

Lying in a rich part of the Milky Way, Cassiopeia contains numerous star clusters. Best is M52, near the border with Cepheus, a fuzzy patch in binoculars that splits into a field of faint stars as seen in a telescope. One star, an 8th-magnitude orange giant, is brighter than the rest. Another cluster visible as a fuzzy patch in binoculars is M103, which telescopes show to consist of a scattering of faint stars with a central diamond shape. Nearby is NGC 663, a large, scattered group of stars of various brightnesses.

Look also at the cluster NGC 457, next to the 5th-magnitude star Phi (φ) Cassiopeiae, which is apparently a true member of the cluster. Since the cluster lies 10,000 light years away, Phi Cas must be an astonishingly luminous supergiant, giving out the light of 200,000 Suns, four times that of Rigel. As seen through a telescope, the stars of NGC 457 appear to be arranged in chains.

Two other objects in Cassiopeia deserve mention, although both are beyond reach of amateur instruments. They are the

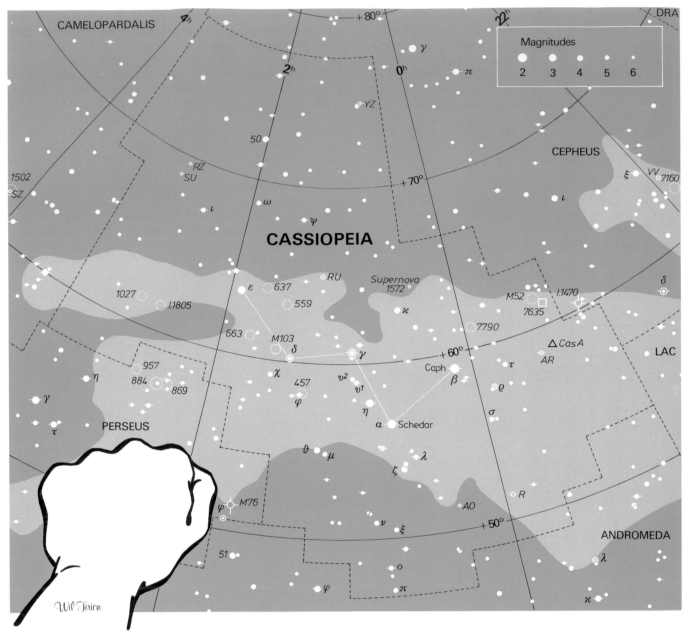

remains of exploded stars, or supernovae. One, known as Tycho's star, flared up near Kappa (κ) Cassiopeiae in 1572. At its peak it became as bright as Venus, and remained visible to the naked eye for sixteen months. It was carefully observed by the great Danish astronomer Tycho Brahe, after whom it is named. All that is left of Tycho's star are some faint wisps of gas, visible only on professional observatory photographs, coinciding with a radio source.

The strongest radio source in the sky, Cassiopeia A, lies almost three degrees south of the cluster M52. Cassiopeia A comprises the remains of another supernova, estimated from its expansion rate to have exploded in the seventeenth century. But it apparently went unseen at the time, possibly because it was abnormally faint. Astronomers have still not solved the mystery of Cassiopeia A.

Cepheus

Cepheus, representing the husband of Cassiopeia, lies close to the north celestial pole. In shape it resembles a squat tower with a steeple. Look first beneath the base of the tower to find Mu (μ) Cephei, one of the most strongly coloured stars in the sky. Sir William Herschel called it the Garnet Star because of its deep red tint, which is notable in binoculars and small telescopes. Mu Cephei is a red supergiant with the luminosity of 50,000 Suns. Like many red supergiants and giants, it is erratically variable, fluctuating between magnitudes 3.4 and 5.1. Take a look at this star next time you are out, and try to estimate its brightness by comparing it with Zeta (ζ) Cephei, magnitude 3.4, Epsilon (ε) Cephei, magnitude 4.2, and Lambda (λ) Cephei, magnitude 5.0.

In the middle of the tower of Cepheus lies Xi (ξ) Cephei, a tidy double star for small telescopes. Its components are of magnitudes 4.4 and 6.5, colours white and yellow. The two stars form a genuine pair, orbiting each other every 4000 years or so. Next try Beta (β) Cephei, a magnitude 3.2 star with an 8th-magnitude companion, difficult to distinguish in the smallest telescopes because of the brightness difference. Beta itself is slightly variable in brightness, by about 0.1 magnitude, virtually undetectable to the naked eye.

Finally we come to the celebrity of the constellation, Delta (δ) Cephei. A small telescope shows it as an easy and attractive double, consisting of a 4th-magnitude yellow primary whose magnitude 6.3 companion, bluish in colour, is wide enough to be picked out in binoculars, if held steadily. But the greatest interest of this star is the primary, a yellow supergiant that varies regularly every 5 days 9 hours. At its brightest, Delta Cephei reaches magnitude 3.6, twice as bright as when at its faintest, magnitude 4.3. These variations can easily be tracked with the naked eye. They are caused by pulsations in the size of the star, like a slowly beating heart. In the case of Delta Cephei, its diameter changes from 32 to 35 times that of the Sun.

The varibility of Delta Cephei was discovered in 1784 by John Goodricke, a 19-year-old amateur astronomer in York, England. Tragically, while continuing his observations of this star, Goodricke caught pneumonia and died, aged 21.

Delta Cephei is the prototype of a class of variable stars known simply as Cepheids, which are of particular importance to astronomers. Cepheids are supergiants that have surface temperatures of 5000 K to 6000 K, similar to that of the Sun, so they appear yellowish in colour. But being much bigger than the Sun, they are also very much brighter (5000 times brighter than the Sun in the case of Delta Cephei), and hence they are easily visible over great distances. Delta Cephei itself is 1300 light years from us.

Cepheids vary in brightness by up to one magnitude (i.e. up to 2.5 times), with periods ranging from about 1 to 70 days. Over 700 Cepheids are now known in our Galaxy (the pole star, Polaris, is one) and they have been discovered in other galaxies too.

The importance of Cepheids to astronomers is that there is a direct link between their inherent brightness (their 'wattage') and their period of variability. The brighter the Cepheid, the longer it takes to vary. By using this relationship, astronomers can deduce the wattage of a given Cepheid simply by monitoring the period of its light changes.

But a star's brightness as seen from Earth is affected by its distance from us. So astronomers can work out how far away a Cepheid is by comparing the star's computed wattage with the brightness at which it appears in the sky. Hence Cepheids serve as marker beacons for measuring distances in space.

Suitable comparison stars for following the light changes of Delta Cephei are Zeta (ζ), Epsilon (ε) and Lambda (λ) Cephei, the same trio as used for Mu Cephei. These stars are shown on the accompanying chart.

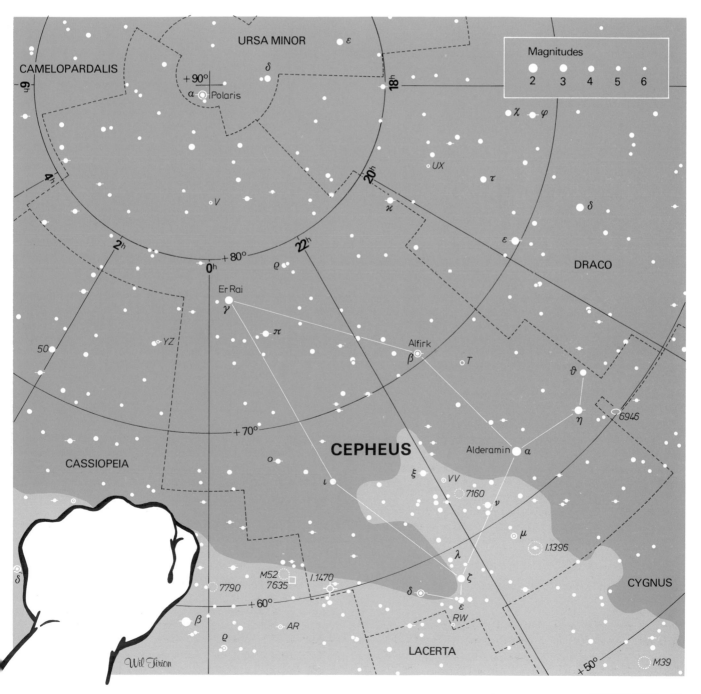

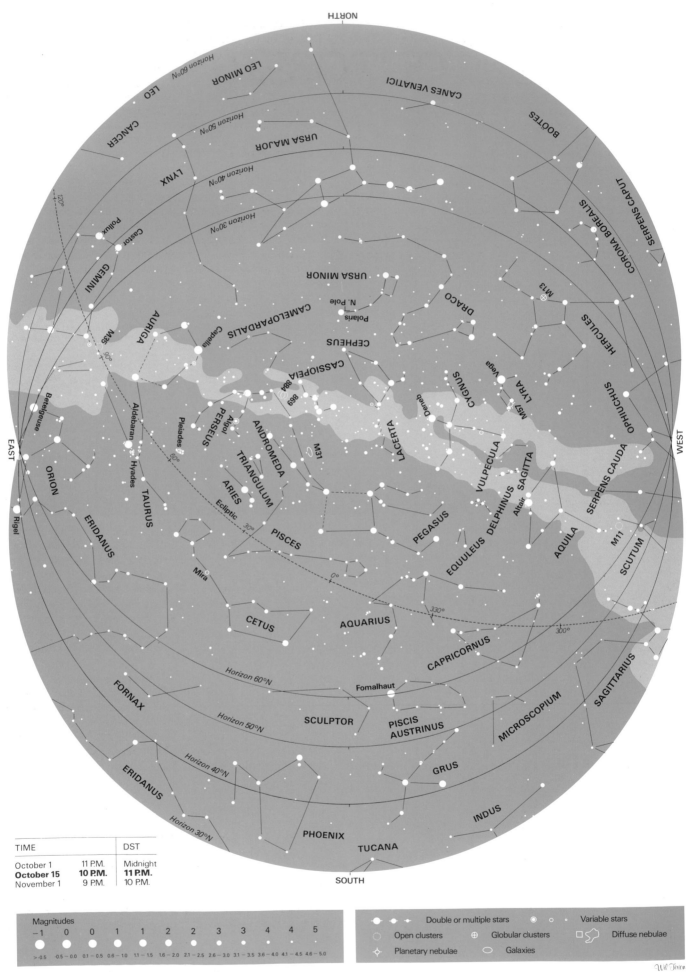

NORTH

TIME		DST
October 1	11 P.M.	Midnight
October 15	**10 P.M.**	**11 P.M.**
November 1	9 P.M.	10 P.M.

SOUTH

Magnitudes

−1	0	0	1	1	2	2	3	3	4	4	5
>-0.5	-0.5 – 0.0	0.1 – 0.5	0.6 – 1.0	1.1 – 1.5	1.6 – 2.0	2.1 – 2.5	2.6 – 3.0	3.1 – 3.5	3.6 – 4.0	4.1 – 4.5	4.6 – 5.0

Double or multiple stars   Variable stars
Open clusters   Globular clusters   Diffuse nebulae
Planetary nebulae   Galaxies

Wil Tirion

OCTOBER

KEY STARS

The Summer Triangle still lingers in the western sky, reluctant to give way to autumn. Due south, the Square of Pegasus stands high while Fomalhaut twinkles over the southern horizon. Between Pegasus and the north celestial pole, the W shape of Cassiopeia is well presented. To the east are Perseus and Auriga, embedded in the Milky Way's starry path. On the eastern horizon, Orion and Gemini are rising. The Plough dips low in the north, and is below the horizon for observers at latitude 30 degrees.

The planets this month

Venus

1988 A morning object at mag. –3.6, moving from Leo into Virgo. Passes Regulus on October 4.

1989 An evening object, moving from Libra across Scorpius and into Ophiuchus, reaching mag. –4.0 by the end of the month. Passes Antares on October 16.

1990 Too close to the Sun for observation throughout the month.

1991 A morning object in Leo, starting the month at mag. –4.3. Passes Regulus on October 8 and Jupiter (mag. –1.4) on October 17. Ends the month near the border with Virgo.

1992 A morning object. Moves through Libra and Scorpius, ending the month in Ophiuchus.

Mars

1988 In Pisces, fading rapidly from mag. –2.5 to –1.7 during the month.

1989 Too close to the Sun for observation throughout the month.

1990 In Taurus, north of Aldebaran, throughout the month, brightening from mag. –0.8 to –1.4.

1991 Too close to the Sun for observation throughout the month.

1992 In Gemini, brightening from mag. 0.4 to 0.1 during the month. Ends the month south of Castor and Pollux, which it outshines.

Jupiter

1988 In Taurus at mag. –2.3, between the Hyades and Pleiades clusters.

1989 In Gemini at mag. –2.0.

1990 In Cancer at mag. –1.6.

1991 In Leo, at mag. –1.4. Overtaken by brilliant Venus on October 17.

1992 Emerges from the morning twilight in the first week of October, at mag. –1.2 in Virgo.

Saturn

1988 In Sagittarius at mag. 0.8.

1989 In Sagittarius at mag. 0.8.

1990 In Sagittarius at mag. 0.7.

1991 In Capricornus at mag. 0.7.

1992 In Capricornus at mag. 0.7.

October meteors

One of the year's less spectacular meteor showers, the Orionids, radiates from near the Orion–Gemini border every October, reaching a peak of about 20 meteors per hour around October 21, although some activity can be seen for up to a week either side of this date. Orionid meteors are fast-moving and can leave dusty trains, but they are not particularly bright so a dark site will be needed to see them well. Like the Eta Aquarids in May, they are dust released from the famous Halley's Comet. The Orionid radiant does not rise until late, so observations are best made after midnight. The radiant lies due south around 4.30 am.

Andromeda

The most famous of all Greek myths, the story of Perseus and Andromeda, is depicted in the skies of autumn. It is the original version of George and the dragon. The tale tells how beautiful Andromeda was chained to a rock by her father, Cepheus, as a sacrifice to appease a terrible monster from the sea that was devastating his country's coastline. Fortunately the hero Perseus happened along in the nick of time, slayed the monster and married Andromeda.

All the characters in this story – Andromeda, her parents Cassiopeia and Cepheus, and her rescuer Perseus – are represented by adjacent constellations. Even the sea monster lies nearby, in the form of the constellation Cetus.

Andromeda itself is not a particularly distinctive constellation. Its most noticeable feature is a curving line of stars that branches off from one corner of the Square of Pegasus, forming an immense dipper shape. In fact, the star at the corner of the Square of Pegasus is Alpha (α) Andromedae. In Greek times this star was considered common to both Andromeda and Pegasus, but now is assigned exclusively to Andromeda. Its two alternative names, Sirrah and Alpheratz, come from the Arabic meaning horse's navel, reflecting its former association with Pegasus.

Andromeda's body can be traced along the line of second-magnitude stars from Alpha, which marks her head, via Beta (β), her waist, to Gamma (γ), her left foot chained to the rock. Gamma Andromedae is well worth a closer look, for it is an impressive double star with a fine colour contrast, like a tighter version of Albireo in Cygnus. The main star of Gamma Andromedae, magnitude 2.2, is an orange gaint very similar to Arcturus. Small telescopes show that it has a 5th-magnitude blue companion. This companion star is itself a close double, but apertures of at least 220 mm are needed to separate the two components.

South of Gamma Andromedae is an even easier double, 56 Andromedae, whose 6th-magnitude components are so wide apart they are divisible in binoculars. This pair of orange gaint suns lies on the edge of NGC 752, a widespread cluster of faint stars, but they are foreground objects, not members of the cluster.

Andromeda contains one of the easiest planetary nebulae for

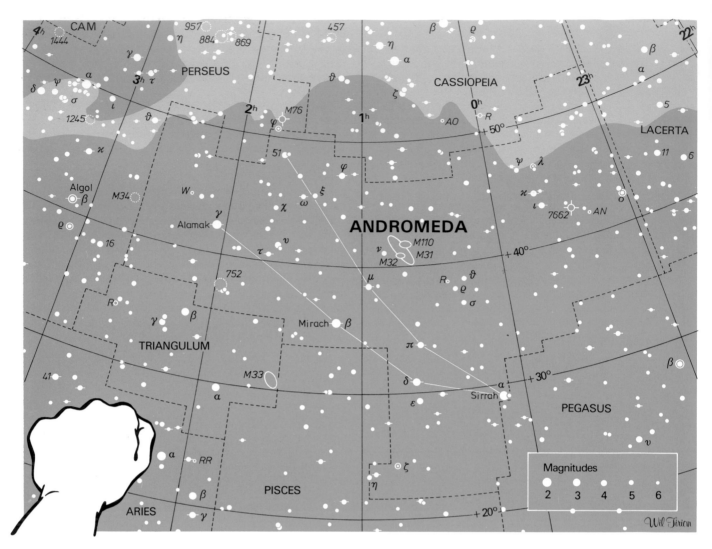

small telescopes, NGC 7662. Look for it one Moon's breadth from the 6th-magnitude star 13 Andromedae. At first glance it resembles a fuzzy star of 9th magnitude. But higher powers show its slightly elliptical outline, similar in size to the disk of Mars, with the luminous blue-green tinge typical of planetary nebulae. Anyone who thinks that planetary nebulae require large apertures should try NGC 7662.

Overshadowing all else in Andromeda is the great spiral galaxy M31, a twin of our own Milky Way. A line from Beta through Mu (μ) Andromedae points to it. Under even averagely clear skies, M31 appears to the naked eye as a faint elliptical glow, high overhead. M31 is the farthest object visible to the naked eye, 2 million light years away. As you gaze upwards at M31, reflect that the light from it entering your eyes tonight has been travelling for 2 million years, since the time when our ape-like ancestors were roaming the plains of Africa.

M31 has a special place in the history of astronomy, for it was the first object to be identified as a separate galaxy outside our own Milky Way. In 1924 the American astronomer Edwin Hubble used the 2.5-m Mount Wilson reflector, then the largest telescope in the world, to take long-exposure photographs of M31. These photographs showed that it contained individual stars, but so faint that they must lie far beyond the stars of our Milky Way. Until then, M31 and objects like it had been generally assumed to be spiral-shaped clouds of gas within our own Milky Way. Now we know that our Milky Way is but one galaxy in a Universe full of countless other galaxies, stretching as far as the largest telescopes can see.

It is easy to understand why previous astronomers failed to recognize the true nature of M31. Even in large telescopes it appears as a nebulous patch with no hint of individual stars. Binoculars show it unmistakably, extending up to several Moon diameters, half the width of your outstretched fist. It appears elliptical in shape because we are seeing it at an angle. See how far you can trace its faint outer regions, for its extent will depend on sky conditions and the instrument used. Small telescopes tend to show only the brightest inner regions of the galaxy.

M31 has two small companion galaxies visible in small telescopes. One Moon diameter south of M31 lies M32, looking like a fuzzy 8th-magnitude star. The other, NGC 205, lies twice as far to the north and is more difficult to spot, being larger and more diffuse.

Seen from outside, our own Galaxy would resemble M31. Of the two, M31 seems to be slightly the larger, with a diameter of about 150,000 light years and containing 300,000 million stars, twice the number in our Milky Way. Do any of those stars have planets on which inquisitive life forms are looking back at us?

Our Galaxy and M31 are the two largest members of a cluster of about three dozen galaxies called the Local Group. To find another member of the Local Group, look over the border from Andromeda into Triangulum. There lies M33, another spiral galaxy, somewhat farther off than M31. It is presented face-on to us, and covers a larger area of sky than the full Moon, but it is notoriously difficult to detect because its light is so pale. Dark skies are essential to see M33, and binoculars are better than telescopes because they condense its thinly spread light. Sometimes observers overlook M33 because it is much larger than they expect.

Pegasus

Pegasus represents the winged horse that was born from the blood of Medusa when she was beheaded by Perseus. Only the front quarters of the horse are depicted in the sky. Pegasus is sometimes identified as the horse of Perseus, but in fact it was another hero, Bellerophon, who rode Pegasus. The constellation is the seventh-largest in the sky, but contains no outstandingly bright stars. Its most distinctive feature is the Great Square whose corners are marked by four stars of second and third magnitudes, although only three of the stars actually belong to Pegasus. The fourth star is Alpha Andromedae, which on some old maps was called Delta Pegasi.

Pegasus is upside-down in the sky, for the Great Square represents his body and the crooked line of stars leading from the bottom right of the Square to Epsilon (ε) Pegasi marks his neck and head. The two lines of stars from the upper right of the Square are his forelegs.

The Square of Pegasus is a large feature, with a diameter of over 15 degrees, or about the width of two outstretched fists, enough to contain a line of more than 30 Moons side by side. Yet within this large area of sky, only a handful of stars are visible to the naked eye. How many stars can you count here?

The star at the top right of the Square, Beta (β) Pegasi, is an immense red giant star, approximately 100 times the diameter of the Sun. It varies irregularly in size and brightness between magnitudes 2.4 and 2.7. Compare it with the other stars of the Square, Alpha (α) Pegasi, magnitude 2.5, and Gamma (γ) Pegasi, magnitude 2.8.

The brightest star in the constellation is Epsilon (ε) Pegasi, otherwise known as Enif, from the Arabic meaning nose. It is a brilliant orange supergiant 500 light years away, appearing of magnitude 2.4. Binoculars or small telescopes show it to have a wide companion star of 8th magnitude.

Above and to the right of Epsilon Pegasi is the greatest treasure of this constellation, the globular cluster M15. It is just too faint to be seen with the naked eye under normal conditions, but shows up clearly in binoculars as a rounded, fuzzy patch of light. A 6th-magnitude star lies nearby, acting as a guide to its location.

Small telescopes show that it has a bright core, where the stars crowd together most densely, although small apertures will be unable to resolve individual stars. The rest of the cluster is like a misty halo around this bright centre. M15 lies 30,000 light years away. It is known to emit X-rays, so possibly it contains a black hole.

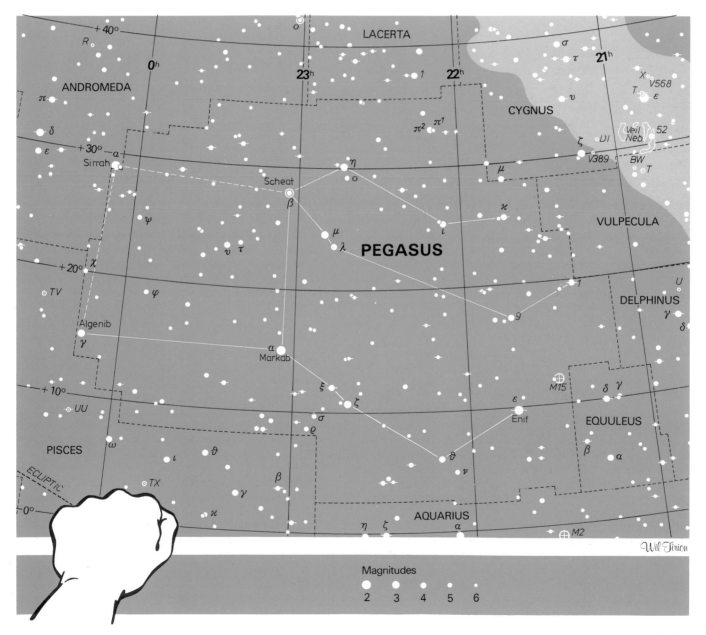

Magnitudes

2 3 4 5 6

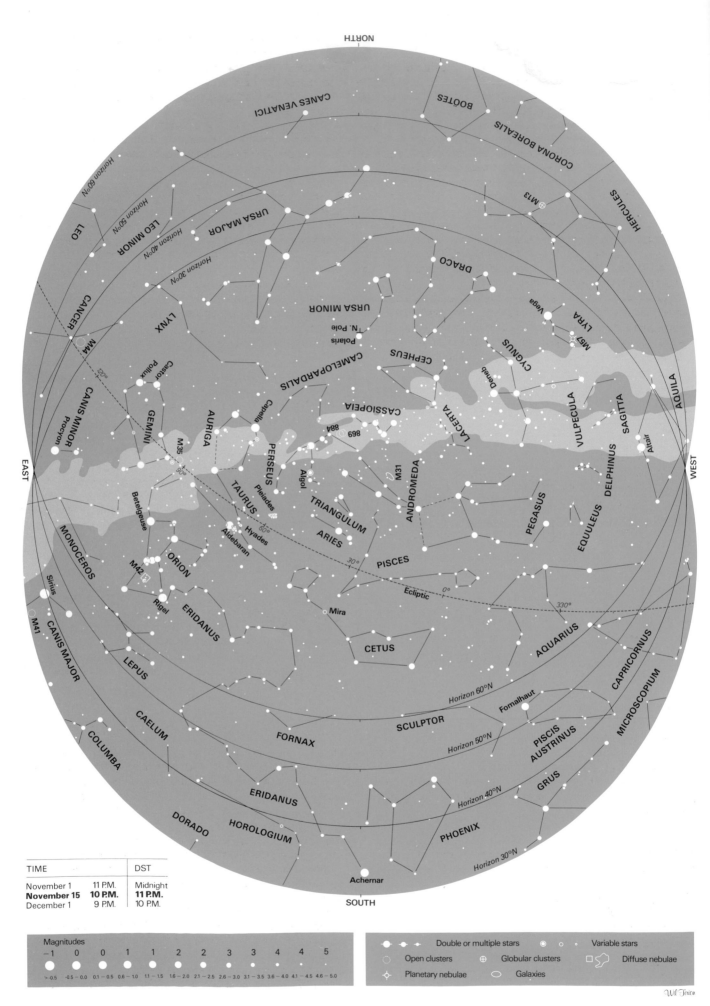

TIME		DST
November 1	11 P.M.	Midnight
November 15	**10 P.M.**	**11 P.M.**
December 1	9 P.M.	10 P.M.

Magnitudes

−1	0	0	1	1	2	2	3	3	4	4	5
>-0.5	-0.5 – 0.0	0.1 – 0.5	0.6 – 1.0	1.1 – 1.5	1.6 – 2.0	2.1 – 2.5	2.6 – 3.0	3.1 – 3.5	3.6 – 4.0	4.1 – 4.5	4.6 – 5.0

Double or multiple stars   Variable stars

Open clusters   Globular clusters   Diffuse nebulae

Planetary nebulae   Galaxies

Wil Tirion

NOVEMBER

KEY STARS

The Milky Way arches from horizon to horizon, passing overhead through Cassiopeia and Perseus. The Summer Triangle reluctantly departs in the west, but the Square of Pegasus is still high in the south-west. Due south is the barren area of Pisces, Cetus, and Eridanus, with no prominent stars, although the first-magnitude star Achernar just peeps above the southern horizon as seen from 30 degrees north. Aldebaran, the red eye of Taurus the bull, and yellow Capella stand prominently in the south-east, followed by Gemini and Orion, signalling the approach of winter.

The planets this month

Venus

1988 A morning object in Virgo at mag. –3.5, passing Spica on November 17. Reaches Libra by the end of the month.
1989 Moves through Sagittarius in the evening sky during the month. Passes Saturn on November 16. Greatest elongation (maximum separation from the Sun) is on November 8, at mag. – 4.1.
1990 Too close to the Sun for observation throughout the month.
1991 A morning object. Starts the month in Leo but soon crosses the border into Virgo, where it remains for the rest of the month. Passes Spica on November 29. At greatest elongation on November 2, mag. –4.1.
1992 In the evening sky at mag. –3.5, moving from Ophiuchus into Sagittarius.

Mars

1988 In Pisces. Fades rapidly during the month from mag. – 1.7 to –0.7 as Earth and Mars move apart.
1989 Moves out of twilight into the morning sky in the second half of the month, as a second-magnitude object in Libra.
1990 In Taurus, north of the Hyades. At opposition (due south at midnight) on November 27 at mag. –1.8, 77 million km from Earth.
1991 Too close to the Sun throughout the month. At conjunction (directly behind the Sun) on November 8.
1992 Starts the month in Gemini, ends up just over the border in Cancer. Brightens rapidly during the month from 0.0 to –0.6 as it approaches Earth.

Jupiter

1988 In Taurus, between the Hyades and Pleiades clusters. At opposition (due south at midnight) on November 23 at mag. – 2.3, 600 million km from Earth.
1989 In Gemini at mag. –2.2.
1990 In Cancer at mag. –1.8.
1991 In Leo at mag. –1.6.
1992 In Virgo at mag. –1.4.

Saturn

1988 In Sagittarius at mag. 0.7.

1989 In Sagittarius at mag. 0.8. Overtaken by Venus (mag. –4.2) on November 16.
1990 In Sagittarius at mag. 0.8.
1991 In Capricornus at mag. 0.8.
1992 In Capricornus at mag. 0.9.

November meteors

Two meteor showers can be seen in November: the Taurids and the Leonids. The Taurids have an extended maximum that lasts for about ten days, from November 3 to November 13 when about twelve meteors an hour may be seen coming from the region near the Hyades and Pleiades clusters. There are two radiants to this stream, one 8 degrees north of the other. Taurid meteors are debris from Comet Encke, the comet with the shortest known orbital period, 3.3 years. Taurids are slow-moving with a high proportion of bright fireballs, which makes the display more impressive than their low numbers might suggest.

Altogether more erratic is the Leonid shower, which radiates from near Gamma (γ) Leonis, peaking on November 17. Leonids are short, stabbing meteors, often flaring at the end of their paths. Many leave persistent trains. Since Leo does not rise until late, observations have to be made after midnight. The Leonids are associated with comet Tempel-Tuttle, which has an orbital period of 33 years. Rates are usually modest, no more than ten an hour. But spectacular Leonid storms can occur at 33-year intervals, when comet Tempel-Tuttle returns to the inner region of the solar system and the Earth encounters a dense swarm of cometary dust. For instance, in 1966 observers in the United States saw Leonid meteors falling from the sky like rain, reaching a peak of thousands a minute. Another spectacular display may be expected when the Leonids' parent comet next returns in 1999.

Perseus

Perseus, a son of Zeus, is the hero in the most enduring of Greek myths, that of Perseus and Andromeda. His adventures began when he was sent to kill Medusa, one of the three Gorgons, winged monsters with snakes for hair who were so terrible to behold that anyone who looked at them was turned to stone – literally petrified. Perseus sneaked up on Medusa while she was sleeping, looking only at her reflection in his shield so that he escaped being petrified. He decapitated Medusa and flew off with her head in a bag. While flying home, he saw the beautiful maiden Andromeda, chained to a rock as a sacrifice to a sea monster. Perseus swooped down, killed the monster, and claimed Andromeda as his bride.

In the sky, Perseus is depicted holding aloft the head of Medusa, whose evil eye is marked by Beta (β) Persei, known by its Arabic name of Algol, meaning the ghoul or demon. The name is certainly appropriate, for Algol is a variable star that appears to 'wink' at regular intervals, from magnitude 2.1 to 3.4. Algol's

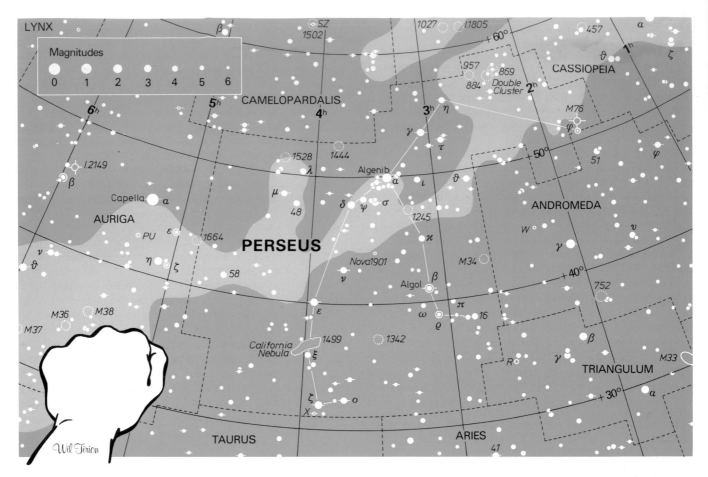

light variations were first recognized by the Italian astronomer Geminiano Montanari in 1667, and the periodicity of the variations was first measured by the English amateur astronomer John Goodricke in 1783. Goodricke also suggested, correctly, that the variations were caused by eclipses.

Algol is an eclipsing binary, consisting of two stars that regularly pass in front of each other as seen from Earth. The stars are too close together to be distinguished individually in a telescope, but analysis of the light from Algol reveals that the brighter of the pair, Algol A, is a white star 100 times as luminous as the Sun. Its companion, Algol B, is a larger but fainter orange star that covers about 80 per cent of Algol A during the eclipses. There is also a third star in the system, Algol C, but this does not participate in the eclipses.

Algol's eclipses occur every 2 days 20 hours 49 minutes. In a mere five hours, Algol fades to one-third of its usual brightness, from magnitude 2.1 to 3.4, then rises again to magnitude 2.1 in another five hours. These variations are easily followed with the naked eye. Compare Algol with Beta (β) Cassiopeiae, magnitude 2.3; Delta (δ) Cassiopeiae, magnitude 2.7; Epsilon (ε) Persei, magnitude 2.9; Delta (δ) Persei, magnitude 3.0; Alpha (α) Trianguli, magnitude 3.4; and Kappa (κ) Persei, magnitude 3.8. You should end up with a light curve looking like the one shown here. A secondary minimum occurs when Algol A eclipses Algol

B, but the fading is so slight that it is undetectable to the eye. Incidentally, do not compare Algol with Rho (ρ) Persei, which is a red giant that fluctuates somewhat irregularly between magnitudes 3.3 and 4.0 every month or two.

Perseus lies in a sparkling part of the Milky Way, rich for sweeping with binoculars. Look at the region around the constellation's brightest star, Alpha (α) Persei, magnitude 1.8. Alpha Persei is a yellow-white supergiant 5000 times more luminous than the Sun, surrounded by a sprinkling of bright stars that form a loose cluster 600 light years away. Note also the cluster M34 on the border with Andromeda, an excellent object for binoculars and small telescopes, consisting of several dozen stars of 8th magnitude and fainter, covering a similar area of sky to the full Moon. M34 lies 1400 light years away.

But the real showpiece of Perseus is the Double Cluster, a twin star cluster on the border with Cassiopeia, marking the hand of Perseus holding aloft his sword. To the naked eye, the clusters resemble a denser knot in the Milky Way. Binoculars and small telescopes show this area to be sown with stars, one of the finest sights in the sky for small instruments.

Each half of the Double Cluster covers a greater area of sky than the full Moon. However, the two clusters are not identical twins. NGC 869, also known as h Persei, is the more star-packed of the two. The other cluster, NGC 884, also known as Chi (χ) Persei, contains some red giant stars, which NGC 869 does not, probably because NGC 884 is older so that the stars in it have evolved more. Both clusters lie in a spiral arm of our Galaxy. NGC 884 is slightly the more distant at 7500 light years away, while NGC 869 is about 300 light years nearer to us. To be visible over such distances the brightest stars in each cluster must be intensely luminous supergiants, similar to Rigel and Betelgeuse in Orion. Binoculars give some idea of how spectacularly large and bright the Double Cluster would appear to the naked eye if it were as close to us as the stars of Orion.

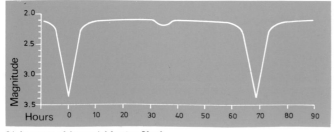

Light curve of the variable star Algol.

Auriga

Next to Perseus in the sky is the hexagonal shape of Auriga, representing a man driving a chariot. By comparison with the flamboyant figure of Perseus, the identity of Auriga is somewhat shadowy. He is usually said to be Erichthonius, a lame king of Athens who invented the four-horse chariot. But another legend identifies Auriga with Myrtilus, charioteer of King Oenomaus and a son of Hermes.

There is no mistaking the constellation's brightest star, Alpha (α) Aurigae, better known as Capella. At magnitude 0.1 Capella is the sixth-brightest star in the sky. Its name comes from the Latin meaning little she-goat, for the charioteer has traditionally been depicted carrying a goat on his left shoulder. Capella is a yellow giant with the light output of 60 Suns, lying 42 light years away. It is the most northerly of all first-magnitude stars.

The stars Eta (η) Aurigae and Zeta (ζ) Aurigae are known as the Kids, the goat's offspring, carried on the charioteer's arm. Zeta is one of two extraordinary eclipsing binary stars in this constellation. It consists of an orange giant over 100 times larger than the Sun, orbited by a blue star similar to Regulus. Although this blue star is smaller than the much brighter giant, it is still about four times the diameter of the Sun. The luminosities of the two stars are 700 and 140 Suns, respectively. Normally Zeta Aurigae shines at magnitude 3.7, but every 2 years 8 months the blue star is eclipsed by the red giant and the magnitude falls to 4.0. The eclipses last six weeks, after which the star returns to its normal brightness.

Even more extraordinary is Epsilon (ε) Aurigae, which has the longest known period of any eclipsing binary, 27 years. The main star is an intensely luminous white supergiant, one of the most powerful stars known, shining with the light of 200,000 Suns and large enough to contain the orbit of the Earth. It lies approximately 4500 light years away. Normally it appears at magnitude 3.0, but every 27 years it is partially eclipsed by a mysterious dark companion. Over a period of four months the star's brightness gradually drops to magnitude 3.8, where it remains for fourteen months before returning to normal.

The nature of Epsilon Aurigae's dark companion was a mystery to astronomers for years. It is now thought to consist of a close pair of stars surrounded by a dark disk of dust that actually causes the eclipses. Epsilon Aurigae was last eclipsed in 1982 to 1984, and the next eclipse will begin late in 2009.

Auriga is notable for an impressive trio of star clusters, M36, M37 and M38, all three being visible in the same field of view through wide-angle binoculars. In binoculars they appear as fuzzy patches, but small telescopes resolve them into individual stars. Each cluster has its own distinct character.

M36 is the smallest and most condensed of the trio, consisting of 60 or so stars of 9th magnitude and fainter, lying 4100 light years away. In binoculars it appears the most prominent of the Auriga clusters. The largest and richest of the Auriga clusters is M37, containing 150 or so stars 4400 light years away. At its centre is a brighter orange star. The most scattered of the clusters is M38, containing about 100 faint stars, 4300 light years away.

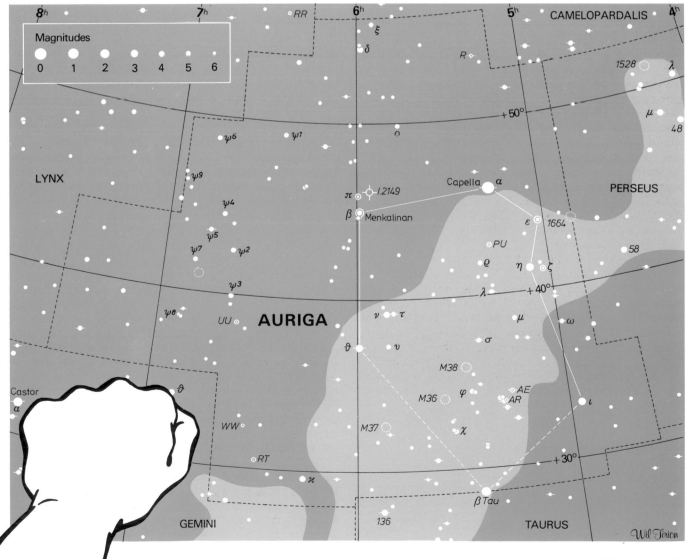

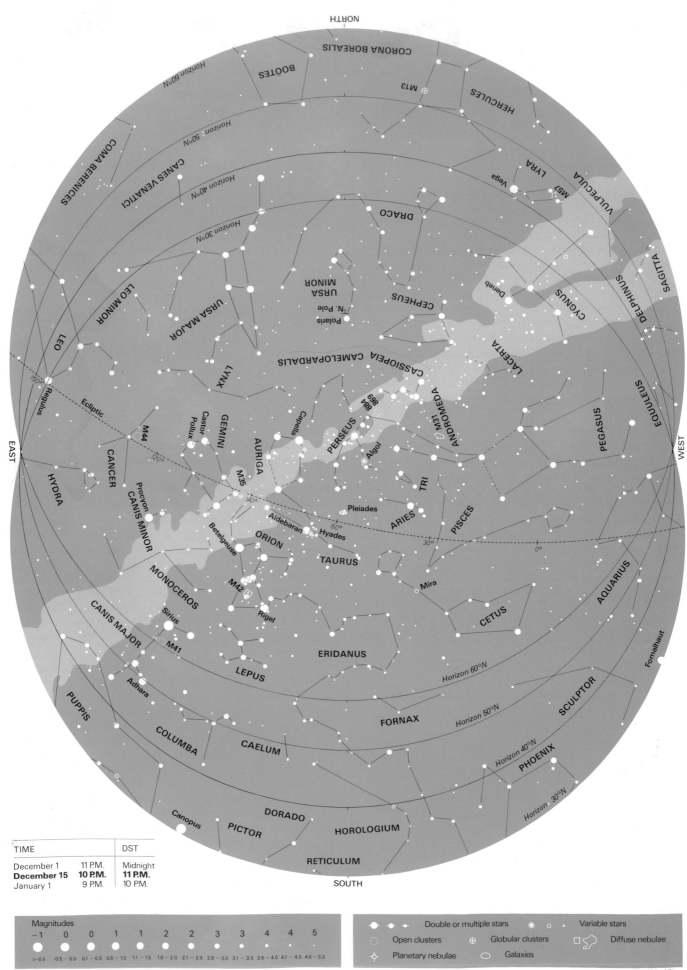

TIME		DST
December 1	11 P.M.	Midnight
December 15	**10 P.M.**	**11 P.M.**
January 1	9 P.M.	10 P.M.

Magnitudes

−1	0	0	1	1	2	2	3	3	4	4	5
> −0.5	−0.5 − 0.0	0.1 − 0.5	0.6 − 1.0	1.1 − 1.5	1.6 − 2.0	2.1 − 2.5	2.6 − 3.0	3.1 − 3.5	3.6 − 4.0	4.1 − 4.5	4.6 − 5.0

Double or multiple stars   Variable stars

Open clusters   Globular clusters   Diffuse nebulae

Planetary nebulae   Galaxies

DECEMBER

The planets this month

Venus

1988 A morning object. Starts the month in Libra, crosses Scorpius and ends the month in Ophiuchus. Mag. –3.4.

1989 Dazzling in the evening sky at mag. –4.4. Moves from Sagittarius into Capricornus.

1990 Too close to the Sun throughout the month.

1991 A morning object. Moves from Virgo into Libra at mag. –3.7, ending the month near the border with Scorpius.

1992 In the evening sky, moving from Sagittarius through Capricornus and ending the month on the border with Aquarius. Brightens from mag. –3.6 to –3.9. Passes south of fainter Saturn on December 22.

Mars

1988 In Pisces, fading rapidly during the month from mag. –0.7 to 0.1.

1989 A second-magnitude object, moving from Libra through Scorpius and into Ophiuchus. Ends the month north of Antares. Mars is mag. 1.8, Antares is 1.0.

1990 In Taurus, near the Pleiades Cluster. Fades rapidly during the month from mag. –1.7 to –0.9.

1991 Too close to the Sun for observation throughout the month.

1992 Moves from Cancer into Gemini, ending the month south of Castor and Pollux. Brightens from mag. –0.7 to an impressive – 1.2.

Jupiter

1988 In Taurus, south of the Pleiades, at mag. –2.3.

1989 In Gemini. Opposition (due south at midnight) is on December 27, at mag. –2.3, when Jupiter is 620 million km from Earth.

1990 In Cancer at mag. –2.0.

1991 In Leo at mag. –1.7.

1992 In Virgo at mag. –1.5.

Saturn

1988 In Sagittarius at mag. 0.7. After the first week of December becomes lost in the evening twilight. Conjunction is on December 26.

1989 In Sagittarius at mag. 0.7. Lost in the evening twilight in the second half of the month.

1990 In Sagittarius at mag. 0.8.

1991 In Capricornus at mag. 0.9.

1992 In Capricornus at mag. 0.9. Brilliant Venus passes less than half a degree south on December 22.

December meteors

The Geminids are one of the brightest and richest showers of the year. At their peak, around December 13–14, as many as 60 meteors an hour may be seen radiating from near Castor. The radiant is well placed for observation all night. Geminid meteors are bright, like the Perseids, but unlike the Perseids they rarely leave trains. Activity falls off sharply after maximum. The Geminids are unique in that their parent body is not a comet, but an asteroid. This asteroid, which has an orbital period of 1.4 years, was discovered in 1983, and was named Phaethon.

Taurus

As the year draws to a close, the familiar figure of Taurus the Bull snorts the night air in the south. His face is marked by the V-shaped cluster of stars called the Hyades, his glinting red eye is the bright star Aldebaran, and his long horns are tipped by the stars Zeta (ζ) and Beta (β) Tauri. In the past, Beta Tauri was regarded as being common to both Auriga and Taurus, and on old maps it was given the alternative designation of Gamma (γ) Aurigae. Now it is exclusively the property of Taurus. Appropriately enough, its Arabic name, El Nath, means the butting one.

Only the top half of the bull is depicted in the sky. This might be explained by one legend, which identifies Taurus with the bull into which Zeus changed himself to seduce Europa. He carried her on his back through the ocean, so perhaps the rest of the bull is immersed beneath the waves. Another identification of Taurus is as the Cretan bull tamed by Hercules as one of his twelve labours. Alternatively, Taurus could represent the Minotaur, the monster half-bull, half-man slain by Theseus.

The brightest star in the constellation, Alpha (α) Tauri, better known as Aldebaran, has a noticeably fiery glow, like a red-hot coal, for it is an orange giant star, 40 times the diameter of the Sun and 100 times as bright. The name Aldebaran comes from the Arabic meaning the follower, from the fact that it follows the Hyades and Pleiades star clusters across the sky. Aldebaran lies 68 light years away, and appears of magnitude 0.9, although it is suspected of slight variability.

Aldebaran looks as though it is a member of the Hyades star cluster, but it is not. It is actually a foreground object at less than half the distance, appearing superimposed on the Hyades by chance.

The Hyades and Pleiades clusters are two of the major tourist attractions in the sky. In mythology, the Hyades and Pleiades were sisters, the daughters of Atlas, but Aethra was mother of the Hyades and Pleione was mother of the Pleiades. The Greeks told that the Hyades nursed the infant Dionysus, a son of Zeus, and were rewarded with a place in the heavens. The name Hyades means the rainy ones, because they were associated with bad weather.

Legend says there were five Hyades, but at least a dozen stars are visible to the naked eye. The Hyades cluster spans 5 degrees of sky, so large that only binoculars can encompass it all. As you scan among the numerous stars that spring into view here, note several bright, wide doubles including Theta (θ) Tauri and Sigma (σ) Tauri. At magnitude 3.4, Theta-2 (θ²) Tauri is the brightest star in the cluster, followed closely by Epsilon (ε) Tauri, magnitude 3.5, and Gamma (γ) Tauri, magnitude 3.6.

Hundreds of stars belong to the Hyades, all of them moving through space together towards a point in the sky near Betelgeuse. Such a moving cluster is of particular importance to astronomers, because the cluster's distance can be accurately derived from the movements of the stars. The Hyades lies 150 light years away, making it the nearest star cluster to us. The distance of the Hyades is the first stepping stone in our scale of the Universe.

But even the Hyades is overshadowed by its near neighbour the Pleiades, popularly known as the Seven Sisters, the most famous of all star clusters. The Pleiades hover like a swarm of flies over the bull's back. To a casual glance, the Pleiades appears as a misty patch, but good eyesight reveals six or seven individual stars, and some people with exceptional eyesight can see ten or more. The misty effect is provided by the dozens of stars in the cluster that are just beyond visibility with the naked eye. Without doubt, the Pleiades cluster is the finest binocular sight in astronomy.

According to myth, the Pleiades were seven nymphs, followers of Artemis, the huntress. Orion, smitten by their beauty, began to pursue them. To save them from his advances, Artemis placed them among the stars. And in the heavens, Orion still forlornly chases the Pleiades across the sky.

Unlike the Hyades cluster, none of whose stars are named, the stars of the Pleiades bear the names of all seven sisters: Alcyone, Asterope, Celaeno, Electra, Maia, Merope and Taygeta. Their father and mother, Atlas and Pleione, are also present. There is even a legend to explain why only six Pleiades, rather than seven, are easily visible. Electra is said to have hidden her face so as not to witness the ruins of Troy, the city founded by her son Dardanus. But some say that the 'missing' Pleiad is Merope, ashamed because she was the only one of the sisters to wed a mortal, Sisyphus. In fact, the naming of the Pleiades has not followed either legend, because the faintest of the named stars in the cluster are Asterope and Celaeno.

In addition to the named stars, binoculars bring into view a glittering array of dozens more, fitting snugly into the field of view. At least 100 stars are members of the cluster, lying just over 400 light years away, nearly three times as far as the Hyades. The main stars of the Pleiades form a shape resembling a squashed version of the Plough or Big Dipper. An interesting demonstration of comparative sizes in the sky is the fact that the full Moon could fit into the bowl of the Pleiades dipper without obscuring either Alcyone or Taygeta. The whole cluster extends across more than two Moon widths.

The brightest member of the Pleiades is Alcyone, also known as Eta (η) Tauri, magnitude 2.9, a blue-white giant 350 times more luminous than the Sun. Pleione, also known as BU Tauri, is an interesting variable of the type known as a shell star. It rotates so quickly, about 100 times faster than our Sun, that it is unstable and throws off shells of gas, causing it to vary gradually in

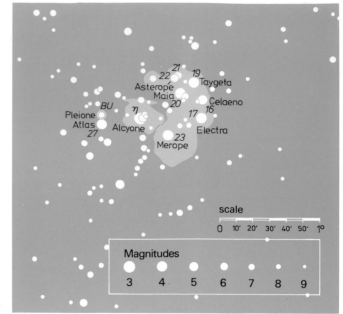

PLEIADES	
Alcyone	mag. 2.9
Atlas	mag. 3.6
Electra	mag. 3.7
Maia	mag. 3.9
Merope	mag. 4.2
Taygeta	mag. 4.3
Pleione	mag. 4.8 – 5.5 (var.)
Celaeno	mag. 5.5
18 Tauri	mag. 5.7
Asterope	mag. 5.8
22 Tauri	mag. 6.4

brightness; currently it is around 5th magnitude. Perhaps Pleione was much brighter in the past, and is really the 'missing' Pleiad.

The Pleiades cluster is relatively young, with an estimated age of 50 million years, compared with 600 million years for the Hyades. But the brightest members of the Pleiades, which stud the binocular field like priceless blue-white diamonds, are no more than a few million years old. Long-exposure photographs show that the Pleiades stars are still surrounded by traces of the cloud of gas and dust from which they formed, but this nebulosity is not visible through amateur instruments under any but the most exceptional conditions.

Taurus is blessed with exceptional objects, for in addition to the Hyades and Pleiades clusters it contains the Crab Nebula, perhaps the most-studied object outside the solar system. The Crab Nebula is the remains of a star much more massive than the Sun that ended its life as a supernova, a cataclysmic nuclear explosion. Astronomers in China saw the star flare up in July 1054. At its brightest it reached magnitude –4, similar to Venus, and was visible in daylight for three weeks. It finally faded from naked-eye view after 21 months, in April 1056. There are no European records of the explosion, probably because astronomy was almost non-existent in Europe at that time.

During its explosion, the Crab supernova must have blazed with the brilliance of 500 million Suns. Twenty-five such supernovae could equal the entire light output of our Galaxy. The outer layers of the shattered star, splashed into space at high

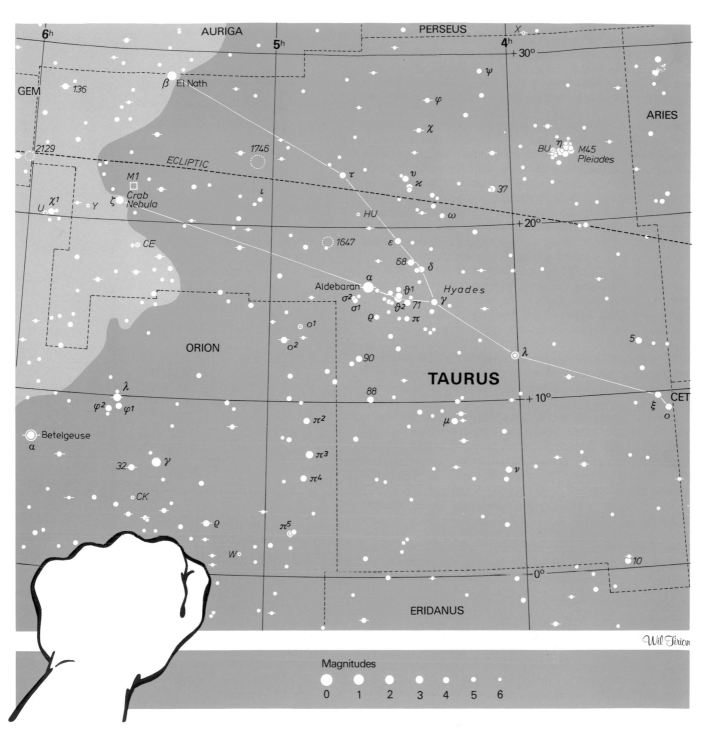

Magnitudes

0 1 2 3 4 5 6

Wil Tirion

speed by the explosion, now form the Crab Nebula, a name due to the nineteenth-century Irish astronomer Lord Rosse who thought that its shape resembled a crab's pincers. Astronomers also know it as M1, the first in the list of nebulous objects compiled by the Frenchman Charles Messier.

The Crab Nebula lies between the bull's horns, just over a degree from Zeta (ζ) Tauri, and 6500 light years from Earth. Under good conditions the Crab Nebula can be glimpsed in binoculars, but for most observers a telescope will be necessary to show its 8th-magnitude elongated misty blur, shaped rather like the flame of a candle. It measures 5 minutes of arc across, one-sixth the Moon's width but six times larger than the disk of Jupiter, so look for an object midway in size between the Moon and a planet. Alas, it is far less spectacular in a telescope than it appears in photographs, but no observer's log is complete without a sighting of the Crab.

At the centre of the Crab Nebula is one of the exceedingly

small and dense objects known as neutron stars, the crushed core of the original star that exploded. A neutron star contains as much mass as the Sun, squeezed into a ball perhaps 20 km across. Being so small, the neutron star can spin very quickly – 30 times a second, in the case of the star in the Crab. Each time it spins, the neutron star emits a flash of light, radio waves and X-rays, like a lighthouse beam. Such a flashing neutron star is termed a pulsar. But, at 16th magnitude, the Crab pulsar is within reach of large telescopes only.

It is a startling fact that, without supernovae, we would not be here. In a supernova, nuclear reactions convert the original star's hydrogen and helium into all the chemical elements of nature, which are then scattered into space, later to be collected up into new stars, planets and, perhaps, life. When you look at the Crab Nebula, recall that the atoms of your own body were produced in such a way, in the supernova explosion of a star that lived and died long before the Sun was born.